SpringerBriefs in Education

We are delighted to announce SpringerBriefs in Education, an innovative product type that combines elements of both journals and books. Briefs present concise summaries of cutting-edge research and practical applications in education. Featuring compact volumes of 50 to 125 pages, the SpringerBriefs in Education allow authors to present their ideas and readers to absorb them with a minimal time investment. Briefs are published as part of Springer's eBook Collection. In addition, Briefs are available for individual print and electronic purchase.

SpringerBriefs in Education cover a broad range of educational fields such as: Science Education, Higher Education, Educational Psychology, Assessment & Evaluation, Language Education, Mathematics Education, Educational Technology, Medical Education and Educational Policy.

SpringerBriefs typically offer an outlet for:

- An introduction to a (sub)field in education summarizing and giving an overview of theories, issues, core concepts and/or key literature in a particular field
- A timely report of state-of-the art analytical techniques and instruments in the field of educational research
- A presentation of core educational concepts
- An overview of a testing and evaluation method
- A snapshot of a hot or emerging topic or policy change
- An in-depth case study
- A literature review
- A report/review study of a survey
- An elaborated thesis

Both solicited and unsolicited manuscripts are considered for publication in the SpringerBriefs in Education series. Potential authors are warmly invited to complete and submit the Briefs Author Proposal form. All projects will be submitted to editorial review by editorial advisors.

SpringerBriefs are characterized by expedited production schedules with the aim for publication 8 to 12 weeks after acceptance and fast, global electronic dissemination through our online platform SpringerLink. The standard concise author contracts guarantee that:

- an individual ISBN is assigned to each manuscript
- each manuscript is copyrighted in the name of the author
- the author retains the right to post the pre-publication version on his/her website or that of his/her institution

More information about this series at http://www.springer.com/series/8914

Cathryn Roos · Gregory Roos

Real Science in Clear English

A Guide to Scientific Writing for the Global
Market

 Springer

Cathryn Roos
Perth, WA, Australia

Gregory Roos
Murdoch University
Perth, WA, Australia

ISSN 2211-1921   ISSN 2211-193X   (electronic)
SpringerBriefs in Education
ISBN 978-981-13-7819-5   ISBN 978-981-13-7820-1   (eBook)
https://doi.org/10.1007/978-981-13-7820-1

This Springer imprint is published by the registered company Springer Nature Singapore Pte Ltd.
The registered company address is: 152 Beach Road, #21-01/04 Gateway East, Singapore 189721, Singapore

Acknowledgements

The authors are extremely grateful to former colleagues Ian McNaught, Dave Maughan-Brown, Dave Ralph, Beth Wiens, Kate Tindal, Linda Marshall and Virgil Williams, who provided expert feedback on the various sections in this book.

The Story of This Book

This book is for English-speaking writers, educators and scientists, who want to better communicate with the burgeoning number of people using English as a second or foreign language. It provides guidelines and tools for conveying complicated information clearly, without affecting the integrity of the subject matter.

Each guideline alerts native- or near-native English writers to the potential language difficulties that their non-native English readers may encounter. Some techniques are suggested to help improve readability. These guidelines can be applied to any material including lecture notes, exam papers, laboratory manuals, textbooks or journal articles.

The Scientist and the Linguist

The story of this book starts with the scientist, an organic chemistry professor, and the linguist, an English-language lecturer, who both worked in English-medium institutions in non-English-speaking countries. Our students and some of our colleagues had to function in a language that was not their mother tongue. This environment attuned us to the issues faced by those who study and conduct their professional lives in a second—and sometimes third or fourth—language.

The governments of these countries established English-medium educational institutions to keep students internationally relevant, while reducing the drain of tuition capital out of the country. Students are often supplied with highly rated textbooks as recommended by their internationally educated faculty. Unfortunately, the writing in most of these textbooks was linguistically and culturally inaccessible.

We reacted to these issues differently according to our professional perspectives. The linguist struggled to teach students the reading skills and technical words they would need to know for their science courses. But the leap from reading excerpts in English class to reading textbooks in science courses was a huge one for many students. The scientist initially urged students to try harder—surely more effort would enable the readers to clarify the written word. But this is similar to speaking LOUDLY and expecting the normal-hearing foreigner suddenly to understand. It just does not work.

Over time, the scientist and the linguist compared perspectives. The scientist was re-educated by the linguist. He was forced to concede that the poor quality of student comprehension is in large part due to the lack of appropriate materials. This includes textbooks, audiovisual lecture and learning aids, and even examination papers.

The scientist and the linguist set out to test whether subject integrity and language accessibility were compatible. Mostly this involved the following sequence:

1. The scientist wrote a paragraph of 'clear' scientific facts.
2. The linguist tried to read it and said: 'But what does this even *mean*?' 'Why do you need to say all that other stuff then?' 'Where's the subject of this sentence?'
3. The scientist explained it.
4. The linguist suggested alternatives.
5. The scientist conceded and then wrote a much clearer paragraph.

When sufficient iterations of this sequence were carried out, they realized that subject integrity and language accessibility were indeed compatible. The scientist was pleased to find that an introductory organic chemistry textbook could be produced using only around 1200 of the most common 2000 English words along with the topic-specific technical words [1].

Our experience led us to produce this book for science and technology professionals who wish to produce material that non-native readers of English can more readily access.

The Guidelines in This Book
We hope that sharing the simple 'secrets' of our process will help others produce a range of science and technology materials that are accessible to readers who struggle with existing mainstream materials. These guidelines are based on:

- extensive professional experience teaching and interacting with non-native readers of English in scientific and academic settings;
- experience preparing multimedia reading and assessment materials for those readers;
- a *counter-perspective* review of research in linguistics, second-language acquisition and academic reading in a second language. Because much of the relevant research is aimed at improving second-language education, we had to take a counter-perspective view to get the information we needed. Instead of 'How can we improve L2 readers' skills?' we sought answers to the question: 'How can we improve the readability of texts for L2 readers?'

The result is this book that integrates experience with theory to provide practical guidelines in improving readability for readers who use English as a second language.

Chapters 1–3 provide background information about the global status of English in science and technology, the position of non-native speakers of English within the global English environment, and the current efforts to make English accessible to those readers.

The key to improving the accessibility of scientific texts is to understand your readers' needs—and to provide for those needs. Chapter 4 provides insight into the experience of reading in a second language. Examples illustrate the issues.

Chapters 5–9 provide guidelines for improving readability, starting at the word and sentence levels. We then discuss how to improve cohesion and text flow, and then finish with suggestions for page layout and formatting. These chapters include plenty of examples to illustrate how to use the guidelines.

Chapter 10 suggests some useful online tools that can help improve and measure readability levels, including screenshots of the tools and examples of how to use them.

This book does not encompass all aspects of good writing. There are several excellent style guides that do that job. One of particular relevance is Anne E. Greene's Writing Science in Plain English [2]. It provides the fundamentals of good scientific writing. However, such guides do not take into account the needs of L2 readers, which are very different from those of general readers.

Perth, WA, Australia
2019

References

1. Roos, G.H.P., and C.L. Roos. 2015. *Organic Chemistry Concepts: An EFL Approach*. Boston: Academic Press.
2. Greene, A.E. 2013. *Writing Science in Plain English*, 136., Chicago guides to Writing, Editing, and Publishing. Chicago: University of Chicago Press.

Terms Used in This Book

General Terms

- **L1 reader**—a reader whose first language or primary language is English. L1 properly refers to a person's first language—French, Spanish, Mandarin, etc—but we use it exclusively in this book to refer to readers whose first language is English.
- **L2 reader**—a reader whose first language or primary language is any language other than English. L2 readers range from beginner to fluent in reading English. For the purposes of this book, we assume an intermediate to advanced fluency range.
- **Text**—any kind of reading material that readers deal with, such as paragraphs, chapters, books, articles, worksheet questions or exams.
- **Science**—all related fields including: physical sciences, technology, engineering, mathematics, medicine, computer science and information technology.

Grammar Terms

To keep this guide comprehensible to everyone, we have minimised grammar jargon. We have used only the following basic grammar terms needed to explain the bits and pieces of a sentence:

- **Subject**—indicates what or who the sentence is about (the agent). It can be singular or plural. It can be a single thing or person, or multiples joined by commas and 'and'. It can be described by several adjectives.
- **Verb**—indicates the action. Sometimes the action is static—e.g. the verb 'be' (is/are, was/were).
- **Object**—receives the benefit of the action. A subject and a verb are usually necessary in a sentence, but sometimes there is no object.
- **Phrase**—a group of words that have a single function within a sentence. It is not a clause or complete sentence. It can be headed by a noun, adjective, adverb, verb or preposition. A phrase cannot stand alone as a sentence.

- **Clause**—a group of words that contains a subject and a verb. It can be a single simple sentence, or it can join with other clauses to form a compound or complex sentence.

In the following examples of simple, compound and complex sentences:

- subjects are single underlined
- verbs are double underlined
- the object is wave underlined
- phrases are indicated with (parentheses)
- clauses are indicated with [square brackets]

Simple: [Pure water boils (at 100 degrees Celsius)].
Compound: [Pure water boils (at 100 degrees Celsius)] but [seawater boils (at higher temperatures)].
Complex: [Seawater boils (at higher temperatures] because [it contains dissolved solids].

Parts of Speech

Each word in a sentence plays a particular role. These are called parts of speech.

Part of speech	Role	Example
verb	action or state	be, have, do sing, dance, carry on
noun	person, place, thing, concept or quality	engineer, ocean, equipment, technology, beauty
pronoun	substitutes a noun	I, you, he/she me, him/her, them
adjective	describes noun	specific, clear, large
adverb	describes verb, adjective or adverb	well, badly, very, carefully
preposition	links noun to other words	in, on, down, up, over, under, before, after
connector	connects words, clauses or sentences	and, but, or because, although, when, that
determiner	limits or determines nouns	a/an, the, some, many
interjection	emotional outburst	hurrah! ouch!

Contents

About the Authors

Cathryn Roos has more than 30 years experience developing curricula and materials for English as a Foreign Language programs in the Middle East, Canada and Japan. She has developed specialized courses as well as learning and testing materials for disciplines such as health and safety, alternative sources of energy, engineering, general science, early childhood education, office skills and computer applications. She has a BA in Anthropology (U. Manitoba), a TESL Certificate (Carleton U.) and a Master of Education (U. Toronto).

Gregory Roos is an organic chemistry professor, researcher and writer with more than 35 years of experience in academic institutions in seven countries. He has developed chemistry courses and learning material for students of various cultural and linguistic backgrounds in South Africa, the Middle East, Taiwan and Australia. In addition to more than 140 research papers, conference contributions and invited lectures, he is co-author of *Organic Chemical Concepts, an EFL Approach* (Academic Press, 2015), and author of the unique research collections *Key Chiral Auxiliary Applications* (Academic Press, 2014) and *Compendium of Chiral Auxiliary Applications* (Academic Press, 2002). He is currently an Adjunct Professor at Murdoch University, Australia engaged in various academic writing projects. e-mail: realsciencecclearenglish@gmail.com

List of Figures

List of Tables

Chapter 1
The Global Impact of English and Science

Contents

Abstract Asian pop groups sing in it. Global investors make money in it. The aerospace industry relies on it—and flies in it. For better or for worse, English has become a modern lingua franca in our increasingly globalised world. From an obscure West Germanic dialect, English has developed and spread over the past 1500 years to become the language of international trade, communications, economic development and innovation. Since the end of World War II, the spread of English has accelerated and displaced French, Italian, German and Russian as the language of science and technology. This chapter provides a brief overview of the global impact of English and its implications for scientific knowledge.

Keywords International English · Global English · English science · English-medium instruction

1.1 Who Uses English?

Despite its global prevalence, English is nowhere near the largest language by number of native speakers. More than 900 million people speak Mandarin and more than 440 million speak Spanish, whereas fewer than 400 million people speak English as a native language [1]. However, as shown in Fig. 1.1, the combination of native (L1) and non-native (L2) speakers of English makes it the most *commonly*

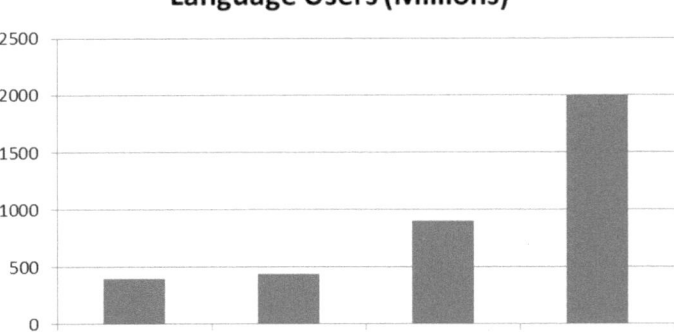

Fig. 1.1 Most popular language usage worldwide

used language in the world. An estimated two billion people across more than 120 nations study or use English to some degree [2].

In addition to countries where the majority of people speak English as a native language, many countries, such as India, Pakistan and Nigeria, use English as an official second language for education and government. For some in those countries, English is, in fact, their native language. In other countries, such as China, Germany and Egypt, English is taught as a foreign language and widely used. Many people are as fluent in English as they are in their native language. In fact, the distinction between native speakers and non-native speakers often blurs so that the labels themselves are controversial.

1.2   What Makes English a Global Language?

The prevalence of English globally has nothing to do with the number of native English speakers in the world. Nor does it have anything to do with the qualities of the language itself. English is not inherently easier or harder to learn than any other language. It might be easy for some, more difficult for others, depending on the amount of their exposure to the language and the degree to which their first language is similar to English.

Instead, the status of English as a global language is a geopolitical phenomenon, beginning with the rise of the British Empire during the industrial era. At its height, the empire covered a quarter of the world including North America, the Caribbean, Australia, New Zealand and much of Africa and South-East Asia. It established English systems of governments and industry that left a powerful linguistic legacy.

But as the British Empire crumbled following World War II, the United States rose in power with its post-war economic boom. With the economic, political and military dominance of the United States, the global impact of English increased exponentially.

1.3   The Synergy of English and Science

From 1995 to 2010, the use of English expanded more than it had in the previous four decades after World War II [2]. This was due to the rapid development of online communication and information dissemination. English and technology became a synergistic force: technological knowledge was spread through English, and English was spread through technology.

However, native speakers of English are no longer the sole drivers of this English–science synergy. For example, of the top ten programming languages, which are all English-based, two were created by non-native English speakers [3].

Now science, the world's driving knowledge base, has a global language. In this role, English has expanded further in scientific achievement than in any other area. So far, English is the only language to retain a true global status over the past 100 years. This, in turn, means that English has assisted the greatest scientific expansion ever recorded. For more than two centuries, science was monopolised by competition and cooperation between a small set of Western nations. Now the further evolution of science has moved into a truly international phase [2].

English is the medium of about 90% of international scientific communication and to participate in global scientific activity means using English. Virtually, all science journals and databases of note are now published in English [4].

The future of science seems inextricably tied to English. English has multiplied the opportunities for collaboration within multinational scientific endeavours. It allows a truly global workforce into corporate research and development, and encourages mobility for academic researchers. This, in turn, allows for an overall better distribution of research talent [2].

1.4   The New 'Owners' of English

With the expansion of English globally, the language has now assumed a life of its own. It is no longer the exclusive preserve of the United Kingdom or the United States. It is a lingua franca whose speakers more often embrace it for practical rather than cultural or ideological reasons. It has also become a communication bridge between different groups of non-native speakers of English [5].

Individuals, as well as national entities, consider English as synonymous with improved incomes, quality of life, business connections and innovation. Indeed, countries whose people have greater English skills tend to produce more researchers and technicians [6]. They have more service-based exports, better internet access, and invest more in research and development, leading to the production of more high-technology output.

Not surprisingly then, the motivation for many people to learn and use English is primarily a practical one. Most people focus on accessing opportunities and improving their futures rather than learning about the Anglo-American culture. Relatively few worry about the prospect of cultural imperialism and homogenisation. Instead, they believe that English facilitates the sharing of ideas in all cultures and between all cultures [7].

This utilitarian approach to English is reflected in the proliferation of focused English courses. ESL/EFL (English as a Second/Foreign Language) programmes have become increasingly specialised. English for Specific Purposes (ESP) programmes provide a range of specialised language courses for specific occupations or professions such as Business English for ESL learners, English for International Tourism or English for Medical Purposes (EMP). English for

Academic Purposes (EAP) focuses on preparing students for academic study. There are even specialised courses that help prepare people for standardised tests such as the Test of English as a Foreign Language (TOEFL) and the International English Language Testing System (IELTS).

1.5 The Expanding Market of English-Medium Instruction (EMI)

In addition to the proliferation of English language programmes (and their acronyms), the number of students travelling to English-speaking countries to access educational opportunities has flourished in recent decades. In fact, many institutions have developed lucrative business models to benefit from the trend. Some have even built satellite branches of their institutions in Asian and Middle Eastern countries [8].

Some students access these opportunities at their own expense. But many are sponsored by their governments seeking to boost their nation's skill level and economy. However, large-scale sponsorship of students is expensive. In addition, some nations are concerned about the cultural impact of students studying abroad. Another issue is that some students end up remaining in their host country. Others return, but with 'bad habits' that may negatively influence local culture. For these reasons, many countries worldwide now encourage or even mandate English as the medium of instruction in local secondary and tertiary science education [9]. As a result, many English-Medium Instruction (EMI) programmes are being established in local institutions in non-English-speaking countries [10].

EMI commonly follows four basic models. A nation's affluence and government policies generally determine the model(s) used.

- branch campuses of academic institutions from English-speaking countries;
- fully fledged local English-medium academic institutions;
- selected courses or programmes delivered in English, but within local language-medium institutions, depending on the availability of appropriate local staff;
- arrangements whereby institutions in English-speaking countries send instructors on a temporary basis to provide courses locally.

The Asia-Pacific region has embraced this movement wholeheartedly [11]. For more than the past two decades, there has been an explosion in higher education, supported by proactive government policies of internationalisation. This has contributed to the more than 8000 EMI courses taught globally in non-English-speaking countries [12].

Asian countries are no longer simply a source of students for the international market. They are also becoming destinations. Universities in China, Malaysia, Japan, South Korea and Singapore now promote themselves as destinations for all nationalities to pursue higher education in the medium of English [13]. It is anticipated that this market will handle in excess of 1.5 million foreign students by around 2022 [14].

These course options appeal directly to more than 5 million international students who currently travel abroad for education. In addition, many students are choosing to study in their own countries in English rather than in their own local language. However, at present, the educational infrastructure in many countries cannot yet support quality EMI. Several reports on the challenges facing effective EMI have appeared [10–12]. Despite these challenges, it seems that the spread of EMI will continue at its current staggering rate. This growth is already competing significantly with traditional study abroad in English-speaking countries such as the United States and the United Kingdom.

1.6   The English Bias

Scientists without a command of English can find themselves severely restricted in access to scientific publications. Although not as marked as it once was, there remains some bias against non-native English speakers [15]. This problem needs to be addressed so that we do not miss contributions from scientists and other innovators who struggle to fully master English as a second language. Currently, this deficiency is addressed either via English editing services or through a collaborator with higher English skills.

Contemporary science and engineering produce between 1.5 and 2 million research papers annually [16]. The United States used to lead publication numbers by a good margin, with China and the United Kingdom filling the next two positions. The number of research articles that contain at least one Chinese author grew by 400% in the decade from 1999. By 2016, in terms of sheer numbers, China had passed the United States [16]. However, even with its high number of publications, Chinese research only receives 4% of global citations in scientific journals. This is well short of the United States with 30% and the United Kingdom with 8%. Because the majority of its journal articles are not in English, Chinese research is less integrated into the global effort. This, however, seems set to change in the near future [16, 17].

In countries with low English proficiency levels, scientists are at a disadvantage in terms of establishing international collaborations. For example, in 2015, only 21% of Chinese scientific research publications involved an international collaborator. By comparison, the Netherlands, Singapore and the Scandinavian countries have more than 50%. This reflects the limit of research access because of the lack of English proficiency [13, 18].

Clearly, current and future students need better access to the English language materials within their chosen fields in order to fully become part of the global scientific community. In the next chapter, we discuss how to help bridge this gap.

References

1. *Ethnologue: Languages of the World*, 21st edn. (2018), http://www.ethnologue.com
2. S.L. Montgomery, *Does Science Need a Global Language? English and the Future of Research* (University of Chicago Press, Chicago, 2013)
3. Education First, *EF English Proficiency Index* (2016) (cited 2018), www.ef.com/epi
4. C. Tardy, The role of English in scientific communication: lingua franca or Tyrannosaurus rex? J. Engl. Acad. Purp. **3**(3), 247–269 (2004)
5. B. Seidlhofer, English as a lingua franca. ELT J. **59**(4), 339–341 (2005)
6. Education First, *EF English Proficiency Index* (2017) (cited 2018), www.ef.com/epi
7. D. Crystal, *English as a Global Language*, 2nd edn. (Cambridge University Press, Cambridge, 2003)
8. P.G. Altbach, J. Knight, The internationalization of higher education: motivations and realities. J. Stud. Int. Educ. **11**(3/4), 290–305 (2007)
9. A. Kirkpatrick, English as a medium of instruction in Asian education (from primary to tertiary): implications for local languages and local scholarship, in *Applied Linguistics Review*, ed. by L. Wei (De Gruyter Mouton, Berlin, 2011), pp. 99–120
10. J. Dearden, *English as a Medium of Instruction—A Growing Global Phenomenon* [Web] (2014), https://www.teachingenglish.org.uk/sites/teacheng/files/pub_E484%20EMI%20-% 20Cover%20option_3%20FINAL_Web.pdf
11. I. Walkinshaw, B. Fenton-Smith, P. Humphreys, EMI issues and challenges in Asia-Pacific higher education: an introduction, in *English Medium Instruction in Higher Education in Asia-Pacific*, ed. by B. Fenton-Smith, P. Humphreys, I. Walkinshaw (Springer, Cham, 2017), pp. 1–18
12. N. Mitchell, *Universities Compete by Teaching in English*. BBC Knowledge Economy Series (2016)
13. A. Kirkpatrick, English as a medium of instruction in East and Southeast Asian Universities, in *Dynamic Ecologies, Multilingual Education*, vol. 9, ed. by N. Murray, P. Humphreys, A. Scarino (Springer Science, Dordrecht, 2014), pp. 15–29
14. T. Eastwood, G. Watson, The increasing pull of China, in *International Education Association of Australia* (IEAA, 2015)
15. D.G. Drubin, D.R. Kellogg, English as the universal language of science: opportunities and challenges. Mol. Biol. Cell **23**, 1399 (2012)
16. The great experiment—can China become a scientific superpower. The Economist, London, 12 Jan 2019
17. Red moon rising—how China could dominate science. The Economist, London, 12 Jan 2019
18. J. Chang, Globalization and English in Chinese higher education. World Englishes **25**(3/4), 513–525 (2006)

Chapter 2
Bridging the Gap

Contents

Abstract Although English has become the lingua franca of science, the degree of access and proficiency varies greatly among individuals and nations. This is one of the downsides of a global language. It may be a few decades before equal access becomes routine (Education First, EF English Proficiency Index (2017) (cited 2018), www.ef.com/epi [1]). Meanwhile, language bias prevents many scientists in the world from fully accessing scientific knowledge. This chapter discusses the issues facing L2 readers and how we can help bridge the gap between L2 readers' proficiency in English and their access to scientific knowledge.

Keywords English-medium instruction · Simplified reading · Simplified English · English proficiency

2.1 English Proficiency Levels

An individual's English proficiency level depends on the following factors:

- *Amount of Exposure to the Language.* Language instruction in schools ranges from a few hours a week to full-time immersion. Learners in English-speaking environments are more likely to learn more and faster, especially if they limit contact with others who speak their native language. Those who learn English in their home countries tend to progress more slowly, even if they are in full-time English programmes. They inevitably interact with their co-learners in their native language, and generally do not use English outside the classroom.

- *Aptitude*. Some learners have an innate ability to acquire languages more easily than others. Intelligence and language aptitude are not necessarily related. Thus, learners may be brilliant scientists but struggle to learn a second language.
- *Level of Literacy in the Native Language*. Learners who have a solid literacy background in their native language are more likely to do well at learning a second language.
- *Motivation*. The degree to which learners acquire a second language may depend on external and internal motivational factors. External motivation might be the degree to which learning another language is critical to one's future. Internal motivation relates to a learner's attitude, confidence and amount of effort applied.

An individual's access to opportunities that promote English proficiency generally depends on the educational policies of their nation—and the nation's ability to pay for them. Annual reports show that English proficiency is highest in European countries and richer Asian countries. For example, Singapore currently lies fifth in the top eight highest proficiency countries. Some other Asian, as well as Latin American countries, are starting to move into the higher skill ranges. Malaysia (13th) and the Philippines (15th) are well placed in the next highest proficiency band. Argentina (25th) and the Dominican Republic (26th) are the highest ranked Latin American countries [1]. Although global trends indicate that access and proficiency are improving overall, some countries still lag behind. Some major global economic players are still lagging, include India (27th), South Korea (30th), China (36th), Japan (37th), Russia (38th) and Taiwan (40th) [1].

2.2 From ABC… to EMI

There are many possible pathways to English proficiency, depending on the opportunities available. The English language education may be provided as:

- a subject in primary and secondary school in which all other subjects are taught in the native language;
- the medium of instruction in primary and/or secondary school;
- one of two languages of instruction in a bilingual programme, along with the native language in primary and/or secondary school;
- a subject in tertiary institutions, as a credit or non-credit option;
- a part-time or full-time course in dedicated private or public language schools;
- a full-time college or university preparatory programme, usually one or more academic years in duration (also known as bridging or intensive English programmes);
- part of a preparatory programme, alongside other skills such as mathematics and study skills;
- on-going individual support in workshops, tutorials and drop-in centres;
- part of a migration settlement programme;
- online courses.

2.3 Proficiency Levels and Standardised English Testing

Most English-medium educational institutions require a certain proficiency of English as measured by a standardised test such as the Test of English as a Foreign Language (TOEFL) or the International English Language Testing System (IELTS). Some institutions administer their own standardised tests. These tests provide scores that range from non-user to expert user [2]. Table 2.1 shows the IELTS and TOEFL scores that correspond to this range of users.

Table 2.1 IELTS and TOEFL English proficiency equivalencies

IELTS description	IELTS score	TOEFL Internet-based test score	TOEFL paper-based test score
Expert user	9	118–120	≥ 645
Very good user	8.5	115–117	626–644
	8	110–114	610–625
Good user	7.5	102–109	581–609
	7	94–101	560–580
Competent user	6.5	79–93	546–559
	6	60–78	530–545
Modest user	5.5	42–59	516–529
	5	35–41	490–515
Limited user	0–4	32–34	450–489
Extremely limited/ intermittent/non-user	0–4	0–31	400–449

Some students with excellent English skills can go directly into their programmes once they have met the normal entrance requirements. However, many must first improve their English skills. Many institutions, whether in English-speaking or in non-English-speaking countries, accept students on a provisional basis and provide preparatory English courses to help them increase their English language skills. In addition, some provide optional or required English courses to provide language support throughout the degree programme.

Each institution requires a minimum score on the standardised English test. This is usually somewhere in the range of 'modest' to 'good user'. But the specific requirement of a programme or institution is based more on supply and demand than on any actual linguistic considerations. In the case of EMIs, local conditions and government policy often influence the requirement. Expediency is also a major factor.

As a result, students who meet the minimum language requirement are not necessarily ready for the linguistic challenges of lectures, textbooks and assessments in English. A 'modest' to 'good' user simply does not have the language skills of a native English speaker counterpart. However, it is a common misconception—among educators as well as students themselves—that achieving the required minimum score suddenly allows a student to function as a native English speaker. It does not. Individual motivation may compensate for a lack of native speaker competence, but not always.

Students at EMI institutions in their own countries face additional problems attaining and maintaining their proficiency in English. Despite being in an English educational environment, they are still surrounded by their own language. Their classmates and sometimes even their instructors speak their native language, making it all too easy to avoid advancing the linguistic skills they need to handle their course material in English.

2.4 Sourcing Suitable Materials

EMI institutions worldwide want the best educational materials for their programmes in order to keep their graduates relevant on the international stage. The majority of these materials are produced in the United Kingdom and the United States, primarily with their own markets in mind. Publishers do produce 'International' editions, but these do not have significantly altered content. Usually, they are simply more cost-effective editions so as to boost sales.

EMI institutions worldwide often choose textbooks and other educational materials because they are widely used in prestigious British or American academic institutions. However, many of these materials fall very short of meeting the needs of global L2 English readers.

As a result, many students struggle to cope with such educational materials, even if they have met the language requirements of the institution. Many become overwhelmed by the assigned readings in standard textbooks. They become discouraged, avoid the assignments and end up performing poorly in their courses even though they may be perfectly able to handle the material intellectually.

In EMI institutions in non-English-speaking countries, instructors recognise this problem and try to bridge the gap by producing study notes for students. Instructors who can speak the students' native language often end up supplementing the materials with explanations in their own language. Both stopgap solutions hinder any further development of students' English proficiency skills. This, in turn, hinders their ability to further pursue their science studies in English.

Another problem arises for students who have previously studied science in their own language and are now in EMI courses. They may have the background science knowledge for a course, but not the English vocabulary. This presents an enormous challenge, especially since the on-going exponential growth of science specialisations spawns ever-expanding lists of specialised terminology. Since most undergraduate science students take a range of introductory science disciplines, preparatory English courses simply cannot adequately prepare students to handle this expanding universe of vocabulary [3, 4].

In addition to the essential technical vocabulary, current educational materials generally use an unnecessarily large vocabulary. Limited studies indicate that students from preparatory English programmes generally acquire a vocabulary of 3,750–4,500 words [5]. Native English speakers have an estimated vocabulary of 15,000–17,000 words [6].

2.5   Materials Fit for Purpose

An increasingly globalised world needs more accessible materials, especially in the increasingly globalised fields of science. EMI institutions around the world need high-quality educational materials that are accessible to L2 readers. But these materials must be fit for purpose.

Materials must have improved readability, but without loss of scientific content. Although current materials contain high-quality topic content, they often lack accessibility to global readers. They contain unnecessarily complex language, distracting features and culturally biased treatment of the information.

International editions of scientific materials must not—and need not—compromise content to be accessible. For example, materials can retain subject-specific technical words, but contain a restricted general vocabulary. This combination of vocabulary, when used with simple and consistent grammatical structures, goes a long way towards making material accessible and readable [7].

References

1. Education First, *EF English Proficiency Index* (2017) (cited 2018), www.ef.com/epi
2. ETS, *Linking TOEFL iBT Scores to IELTS Scores—A Research Report* (2010), pp. 1–17
3. M.F. Ruiz-Garrido, J.C. Palmer-Silveira, I. Fortanet-Gomez (eds.), in *English for Professional and Academic Purposes*, ed. by W. Herrlitz, P van den Hoven. Utrecht Studies in Language and Communication, vol. 22 (Rodopi, Amsterdam, 2010), p. 237
4. D.C. Castano, *A Software/Design Method for Predicting Readability for ESL Students*. English Philology, Complutense University of Madrid, Masters, P.T. Caller (2010)
5. J. Milton, The development of vocabulary breadth across the CEFR levels, in *Communicative Proficiency and Linguistic Development: Intersections Between SLA and Language Testing Research*, ed. by I. Bartning, M. Martin, I. Vedler (European Second Language Association, 2010), pp. 211–232
6. P. Nation, R. Waring, Vocabulary size, text coverage and word lists, in *Vocabulary: Description, Acquisition and Pedagogy*, ed. by N. Schmitt, M. McCarthy (Cambridge University Press, Cambridge, 1997), pp. 6–19
7. G.H.P. Roos, C.L. Roos, *Organic Chemistry Concepts: An EFL Approach* (Academic Press, Boston, 2015)

Chapter 3
Readability

Contents

Abstract We can increase the accessibility of scientific writing to L2 readers by improving readability of the texts. Existing concepts and guidelines provide some help with this, but fall short when applied to communicating science topics to L2 readers. This chapter discusses readability, as well as other attempts at making English accessible to global readers.

Keywords Simplified English · Improving readability · Global English · Science readability · L2 readability

3.1 What Is Readability?

Readability refers to the level of difficulty of a text and how that affects a reader's comprehension, reading speed and level of interest in the material [1]. The aspects of a text that affect readability include:

- vocabulary;
- grammar;
- sentence structure;
- discourse complexity, i.e. the way in which sentences fit together to present arguments and connect ideas coherently;
- supporting graphics and illustrations;
- layout, font and other design features.

Research supports the common sense assumption that controlling the difficulty and/or complexity of these factors increases readability for L2 readers [2].

However, readability is not just about the text and what is on the page. Reader factors also determine readability, and it is important to keep these factors in mind when selecting content. These factors include readers':

- previous knowledge and expertise;
- social, cultural and educational background;
- interests and motivation to succeed.

Increasing readability does *not* mean dumbing down content. Despite the complexity of scientific concepts, data and analysis, scientific texts *can* be made more readable. And it can be done without oversimplifying the content and without giving up key technical vocabulary [3]. Complexity of thought does not have to lead to incomprehensible texts. In fact, the process of making scientific texts more readable usually results in clarifying the concepts.

3.2 Why Improve Readability?

By writing in a way that absorbs a minimum of your readers' mental energy, you can free them up to focus more on the content of your text. Thus, your task is not merely to present information and concepts. It is to make that information understood by removing roadblocks to reading [3]. The goal is to make it harder for L2 readers to misunderstand the text—and therefore make it easier to understand.

With more readable texts, L2 readers can move through material more automatically. They can read faster and better. The faster they read, the less likely they will become discouraged, and the less likely they will abandon the process.

Even very good readers will benefit from more readable texts. Less focus on the reading process means more focus on content, more mental energy with which to absorb complicated scientific material.

Native English speakers are not automatically the best at writing readable texts. In fact, they are often notoriously poor communicators in international environments [4]. In this increasingly English-dominated world, they are more likely to be monolingual and less likely to understand language variations. Some may not even see the need to accommodate or adapt to others. But as we saw in Chap. 1, it is more important than ever to learn to communicate well in an international environment.

3.3 Some Existing Guidelines

A number of guidelines, style guides and concepts have emerged to try to facilitate international communication. Each has its merits, but falls short in some way when dealing with scientific writing for L2 readers.

3.3.1 Plain English

Plain English is a style of communication that emphasises clarity, brevity and avoidance of technical language. It is primarily intended for L1 readers in a general, non-academic context. Plain English is commonly used for government and business communication. Although some aspects of this style are useful, it does not serve our purposes because:

- technical language is essential for readers of science;
- the needs of L2 readers are very different from those of L1 readers. For example, Plain English guides recommend writing as one would speak, using shorter more informal words. This works well for L1 readers, who have acquired a vast array of such language before they have even learned to read or write. But many L2 readers have not acquired much informal language prior to learning to read in English. Thus, what is simple and familiar vocabulary to an L1 reader is not necessarily so to an L2 reader [5, 6].

3.3.2 Basic English

Basic English is a controlled language developed by a linguist in the 1930s to provide the world with a means of global communication [7]. It allows the use of only 850 selected words. The grammar is simplified and restricted, sometimes to the point of being unnatural in normal English.

Basic English was originally intended for normal everyday communication. By adding another 100 words for general science and 50 for any particular science, it was claimed that Basic English would allow international communication in any scientific forum. Even if this had been true in the 1930s, it is highly unlikely in today's world. In any case, Basic English was never widely adopted.

3.3.3 Special English

Special English is a controlled language that avoids idioms and uses a core vocabulary of 1500 words. Voice of America uses it for broadcasting, but it is otherwise not widely applicable.

3.3.4 Simplified Technical English

Simplified Technical English (ASD STE-100) is a trademarked name for a controlled language developed in the 1980s by the aerospace industry for the preparation of international technical documentation. The Aerospace and Defence Industries Association of Europe (ASD) regularly updates the guidelines using feedback from its users [8, 9]. Other industries now use it for a variety of purposes.

The aim of STE is to provide technical writers with guidelines for writing in a clear, simple and unambiguous way so that L2 aerospace workers can easily understand maintenance manuals.

STE consists of a set of writing rules and a dictionary of controlled vocabulary deemed sufficient for any technical sentence. Most of these words are restricted to just one meaning. Most words can only be used in one form, such as a verb or a noun. STE does not provide guidelines for formatting issues, such as typeface, numbering and layout. The focus is on expressing the content.

The STE specifications provide an excellent writing guide for any procedural or laboratory manuals. However, the rules and examples are narrowly focused on instructions. They do not apply readily to other types of writing.

3.3.5 Aviation English

Aviation English is for the specific purpose of facilitating radiotelephony communication between pilots and air traffic controllers [10]. It consists of standard phraseology and restricted grammar and vocabulary to be used in specified ways in routine exchanges.

3.3.6 Seaspeak

Seaspeak is an English-based language for the specific purpose of facilitating communication among maritime personnel who do not share a native language. Like Aviation English, it consists of standard phrases for routine communication situations. It has a limited vocabulary supplemented with foreign words where no suitable English ones exist [11].

3.3.7 Globish

Globish is the trademarked name for a limited form of English developed for international business in the 1990s and early 2000s by a French computer engineer. Inspired by the observation that non-native speakers could communicate better with each other than with native speakers of English, he developed a system of 'correct English without the culture'. It consists of a list of 1500 words, simple sentence structures and a set of rules, such as 'No idioms and no jokes' [12].

3.3.8 International/Global English

International English is an idealised concept rather than a controlled language or specified standard. It refers to the idea that English is used as a means of global communication. The concept acknowledges the difference between an international variety of English and local varieties such as British, American and Australian.

One linguist's vision is that International English will be one of two standard Englishes that all native speakers need to acquire—one for local and national communication and the other for international communication [13].

International English also refers to the desire for an international standard of the language. However, it is unclear how this standard would come about. Unlike other languages such as Spanish and French, English does not have an official body that regulates language standards. Therefore, there are no specific guidelines.

There are some style manuals that provide guidelines for communicating in International English in certain fields, such as:

- The Elements of International English Style [14] for business, marketing and commerce.
- The Global English Style Guide [15] for writing technical documentation.

These manuals provide some useful information, but are limited to their specific subject areas.

3.3.9 English as a Lingua Franca

English as a lingua franca (ELF) refers to the use of English as a common language among predominantly non-native speakers of various languages. Like International English, it is a concept rather than a controlled language or specified standard. The main difference between ELF and International English is the focus. ELF focuses on getting the message across rather than adhering strictly to the native speaker rules of the language.

The concept of ELF is also a recognition that English, as a result of its international use, is shaped as much by its non-native speakers as by its native speakers [16]. Thus, the ownership of the language shifts from native speakers to multilingual, multicultural users. It suggests diversity rather than imperialism. Native speakers of English are expected to be more tolerant and understanding when they communicate with non-native speakers.

As with International English, there are no specific guidelines. Correct English grammar is not a requirement as long as the message is clear.

3.4 Real Science in Clear English

The existing guidelines for facilitating international communication do not recognise the exigencies of scientific writing. Overly specific writing rules and restrictions are not easily applied across all branches of science, technology, engineering and mathematics.

The guidelines in this book recognise that scientific writers must make choices based on their own fields. In other words, it *is* possible to combine real science with clear English [17, 18]. Scientific writers can:

- retain key technical vocabulary while choosing simpler options for all other vocabulary;
- maintain standard English grammar while using language features that do not block L2 readers' comprehension;
- remove or minimise distractions without sacrificing content.

We all want our readers to understand exactly what we mean. But all communication involves some risk of misunderstanding, a risk that increases with communication between people who do not share a first language.

The best way to deal with the risks of miscommunication is to consider the background of the reader—and use that information to write in a way that the reader is most likely to understand. For this reason, the next chapter prefaces our guidelines with a thorough description of the challenges faced by L2 readers.

References

1. K. Collins-Thompson, Computational assessment of text readability: a survey of current and future research. Int. J. Appl. Linguist. **165**(2), 97–135 (2014)
2. S.A. Crossley, D.B. Allen, D.S. Mcnamara, Text readability and intuitive simplification: a comparison of readability formulas. Read. Foreign Lang. **23**(1), 84–101 (2011)
3. D.G. Gopen, J.A. Swan, The science of scientific writing. Am. Sci. **78**, 550–558 (1990)
4. L. Morrison, Native speakers are the world's worst communicators (2016), http://www.bbc.com/capital/41
5. Plain English Campaign (2015), https://www.plainenglish.co.uk/
6. Plain language makes it easier for the public to read, understand, and use government communications (2018), https://www.plainlanguage.gov/
7. C.K. Ogden, *Ogden's Basic English* (1996). 24/03/12 (cited 10 March 2018); 1. http://ogden.basic-english.org/basiceng.html
8. ASD, *The ASD-STE100 Specification* (2017) (cited June 2018), http://asd-ste100.org/about.html
9. ASD, *The ASD-STE100 Software* (2017) (cited June 2018), http://www.asd-ste100.org/software.html
10. J.C. Alderson, Air safety, language assessment policy, and policy implementation: the case of aviation English. Ann. Rev. Appl. Linguist. **29**, 168–187 (2009)
11. International Maritime Organization, *IMO Standard Marine Communication Phrases (SMCP)*, vol. 46/9. Rijeka College of Maritime Studies (Rijeka, 2000), pp. 1–103
12. J.-P. Nerriere, *Globish* (2004), http://globish.com/?&lang=en_utf8

13. D. Crystal, *The English Language: A Guided Tour of the Language*, 2nd edn. (Penguin, London, 1988)
14. E.H. Weiss, *The Elements of International English Style* (M. E. Sharpe, New York, 2005)
15. J.R. Khol, *The Global English Style Guide: Writing Clear, Translatable Documentation for a Global Market* (SAS Institute Inc., Cary, 2008)
16. B. Seidlhofer, English as a lingua franca. ELT J. **59**(4), 339–341 (2005)
17. G.H.P. Roos, C.L. Roos, *Organic Chemistry Concepts: An EFL Approach* (Academic Press, Boston, 2015)
18. G.H.P. Roos, C.L. Roos, Real science in clear English, in *IAFOR: Asian Conference on Education* (IAFOR, Kyoto, 2015)

Chapter 4
The L2 Reading Experience

Contents

Abstract What problems do L2 readers face? An awareness of these problems is the key to writing more readable texts for L2 readers. This chapter discusses a range of factors that affect the L2 reading process.

Keywords Reading process · STEM vocabulary · Science reading · STEM reading for L2 · Reading comprehension · ESL reading

4.1 The Reading Process

When we read—in a first or a second language—we derive meaning via various mental processes. One common way of conceptualising these processes is bottom-up and top-down [1].

Bottom-up processing refers to the way we make sense of the language on a page by starting with the smallest data on the page—the letters or characters—and working upwards to syllables, words, phrases, sentences and paragraphs. At first, this decoding process is slow and inefficient. But with time and practice, the process becomes more automatic, and we get better at extracting meaning from the text. Then, we begin to read more quickly.

Top-down processing refers to the way we make sense of a text starting with the 'big picture' and working down from that using predictions based on our background knowledge. This background can include:

- what we already know about the topic;
- how interested we are in the topic;
- what sociocultural and emotional perspectives we bring to the topic.

As such, reading is not merely a process of absorbing information. It is an interactive cognitive process. We connect what we are reading to concepts already stored in our memories. Good readers can efficiently combine new information with their previous knowledge of the world.

Bottom-up processing is like viewing something through a microscope. Top-down processing is like viewing a landscape from an airplane.

Readers who are familiar with the information in a text can rely less on bottom-up processing. But readers who encounter a lot of new information must rely more on bottom-up processing.

L2 readers tend to rely more heavily on bottom-up processing. This is especially true for students in English-medium academic courses because:

- they are reading mostly unknown material, even if that material builds on previous knowledge;
- they sometimes encounter texts that are outside their cultural or personal experience;
- they lack confidence in using their previous background, even if the material is familiar.

Relying too much on bottom-up processing slows readers down. It is inefficient and frustrating, and it could cause readers to give up. Relying on bottom-up processing also distracts L2 readers from using important top-down skills. For example, they may not be thinking critically about what they are reading. Or they may not be making sense of the text as a whole. Without some top-down processing, readers cannot see the wood for the trees.

4.2 L1 Versus L2 Reading

L2 readers come to the task of reading in English very differently from L1 readers. Table 4.1 outlines these differences.

The well-developed reading techniques and conceptual skills that L2 readers already have may help or hinder their efforts to read in their L2. Some of the techniques that L2 readers bring to the task may interfere with reading in English. Others could be useful. Each language has its own set of optimal techniques. The

Table 4.1 Differences between L1 and L2 readers

L1 readers	L2 readers
• learn language orally before beginning to read	• usually begin to read almost as soon as they begin to learn listening and speaking skills
• have had exposure to four to six years of naturally occurring oral language at home before they begin learning to read in school	• usually do not have several years of exposure to naturally occurring L2 oral language prior to beginning to read in L2
• implicitly understand grammar and have extensive oral vocabulary before starting to read—e.g. for English-speaking children 5,000–8,000 words	• lack extensive L2 vocabulary and grammar when beginning to read
• do not begin learning to read with a preset notion of certain techniques	• already have well-developed reading techniques in their own language
• develop cultural and conceptual abilities simultaneously with language skills	• already have well-developed cultural and conceptual skills in own language

amount of overlap between the optimal techniques of a first and second language depends on:

• how much the L2 reader's language and culture are different from English;
• how much English grammar and vocabulary the L2 reader has learned. L2 readers can only apply their L1 reading techniques to L2 reading once they have learned a critical amount of L2 grammar and vocabulary [2].

4.3   Effects of First Language on L2 Reading

Languages differ in many ways and to varying degrees. In general, the greater the differences between a reader's first and second language, the more challenges the reader must overcome. For example, a Chinese L2 reader must first master an entire unfamiliar alphabet before learning to read English. A French or Spanish reader already knows the alphabet. In fact, those readers will know many English words that have Latin roots.

Differences between languages can result in interference, which slows L2 readers down. If L1 and L2 are similar, L2 readers can more readily transfer their skills.

Reading in a second language can be affected by differences in *sound systems, writing systems, vocabulary, grammar patterns, text cohesion and text organisation.*

4.3.1   Sound Systems

It might seem that sound systems are not at all related to reading. But readers in all languages use their sound system to process text. Even Chinese children, who learn to read by deciphering a set of characters rather than an alphabet, use their sound system to process texts [3].

Many L2 readers do not have a strong oral background in English before learning to read in English. This will limit their ability to use the sound system to help them process texts.

Differences in sound systems can also create challenges for L2 readers. The number of phonemes in different languages varies from as few as 11 to as many as 141. A phoneme is a single distinctive sound, a minimal unit in the sound system of a language. English has 39–44 phonemes, depending on dialect [4]. For example, some languages have fewer phonemes than English. In this case, L2 readers might struggle to distinguish between unfamiliar sounds such as /l/ and /r/. This can mean mixing up words, such as 'read' and 'lead', when reading.

4.3.2   Writing Systems

Writing systems in other languages range from closely related to completely different from English. They could be based on:

- the Roman alphabet, as is English, such as Spanish, French and Italian;
- other alphabets, such as Korean, Russian or Greek. Each letter corresponds (more or less) to a sound in the language;
- a syllabic alphabet. Each letter of the alphabet corresponds to a syllable in the language;
- characters, such as Chinese and Japanese kanji. Chinese characters indicate both a meaning and a sound. Most words are two or more characters. As children learn to read characters, their brains develop a particular set of neural pathways. These pathways are significantly different from those in English. Thus, some L2 readers may need to use their brains in completely new ways in order to make sense of the written word in English.

Of all the alphabet languages, English is the least regular and consistent. It does not have a clear one-to-one relationship between its alphabet and its sounds, as do Finnish and Spanish. This makes it more difficult for L2 readers to recognise the written forms of words they have heard [5].

4.3.3   Vocabulary

Vocabulary knowledge is an important factor in L2 reading comprehension [6]. L2 readers with a larger vocabulary base are more likely to have better reading comprehension skills. But it is not just the size of vocabulary that matters. Better reading comprehension comes from knowing:

- the various meanings of a word;
- its various forms (e.g. happy, happiness, happily);
- how it behaves grammatically in a sentence;
- its use in common phrases [7].

Most research suggests that readers need to know 95–98% of the words in an academic text to understand it [8]. There is increasing evidence that 98% is best. Without that coverage, L2 readers must break the flow of their reading to look up words in the dictionary. This slows down the reading process. It leads to lower comprehension, greater frustration and a higher chance of giving up.

Example 4.1 shows what it is like to read when you know only **80%** of the words. The missing 20% are the most difficult words in this passage.

Example 4.1

Medicinal   is the study of how   drugs can be designed and developed.

This process is helped immeasurably by a detailed understanding of the

and   of the   that are present in the body. The major drug

          are normally large   (   ) such as   and   .

Knowing the   , properties and   of these   is   if we

are to design new drugs. There are a   of reasons for this. First, it is important

to know what   different   have in the body and whether

them is likely to have a   effect in treating a particular disease. There is no

point designing a drug to   a   if one is looking for a new   .

Example 4.2 shows what it is like to read when you know **95%** of the words. The missing 5% are the most difficult words in this passage.

Example 4.2

Medicinal chemistry is the study of how novel drugs can be designed and developed. This process is helped immeasurably by a detailed understanding of the structure and function of the molecular targets that are present in the body. The major drug targets are normally large molecules (   ) such as proteins and   acids. Knowing the structures, properties and functions of these
              is crucial if we are to design new drugs. There are a variety of reasons for this. First, it is important to know what functions different   have in the body and whether targeting them is likely to have a beneficial effect in treating a particular disease. There is no point designing a drug to inhibit a digestive   if one is looking for a new   .

Example 4.3 shows what it is like to read when you know **98%** of the words. The missing 2% is one technical word, macromolecules, that occurs three times.

Example 4.3

Medicinal chemistry is the study of how novel drugs can be designed and developed. This process is helped immeasurably by a detailed understanding of the structure and function of the molecular targets that are present in the body. The major drug targets are normally large molecules (   ) such as proteins and nucleic acids. Knowing the structures, properties and functions of these   is crucial if we are to design new drugs. There are a variety of reasons for this. First, it is important to know what functions different   have in the body and whether targeting them is likely to have a beneficial effect in treating a particular disease. There is no point designing a drug to inhibit a digestive enzyme if one is looking for a new analgesic.

Example 4.4 shows **100%** of the full text.

Example 4.4

Medicinal chemistry is the study of how novel drugs can be designed and developed. This process is helped immeasurably by a detailed understanding of the structure and function of the molecular targets that are present in the body. The major drug targets are normally large molecules (macromolecules) such as proteins and nucleic acids. Knowing the structures, properties and functions of these macromolecules is crucial if we are to design new drugs. There are a variety of reasons for this. First, it is important to know what functions different macromolecules have in the body and whether targeting them is likely to have a beneficial effect in treating a particular disease. There is no point designing a drug to inhibit a digestive enzyme if one is looking for a new analgesic.

Examples 4.1–4.4 give us a good indication of why linguistic researchers are finding that 98% coverage is more desirable. To get 98% coverage of a general academic text, L2 readers need to know between 8,000 and 9,000 word families [9]. Our own analysis of popular first-year university science textbooks indicates that this level of vocabulary is sufficient for physics texts. But earth science commonly requires up to 12,000, chemistry 13,000 and biology more than 18,000 word families.

However, L2 graduates from university English preparatory programmes are not likely to have such a large vocabulary. For example, Cambridge Advanced English graduates—equivalent to 6.5–7.5 IELTS level—have a vocabulary of 3750–4500 words [10]. This falls considerably short of the required vocabulary level for 98% coverage.

Factors that affect L2 readers' ability to read and understand vocabulary in texts are frequency, idiomatic usage, multiple meanings, different pronunciation or spelling, lack of equivalent concepts and lack of similarity.

Frequency

L2 readers are less likely to know words that do not appear frequently in texts. This is because they do not encounter the words often enough to commit them to memory. Research shows that the more times L2 learners encounter a new word, the more likely they are to learn it [8]. There is no evidence that shorter words are easier to acquire than longer words. But higher frequency words tend to be short.

Idiomatic Usage

An idiom is a phrase that cannot be easily understood by knowing the individual words in the phrase. In English, idioms can be phrasal verbs, such as 'give up' or 'bear down on'. Or, they can be more colourful expressions, such as 'to be poles apart', 'counting your chickens before they hatch' and 'kick the bucket'. What may seem a normal, everyday part of language can be very problematic for L2 readers [11].

Although the components of idiomatic expressions may be high-frequency words, the expression in which they are used may not be. Therefore, L2 readers must learn these expressions as a holistic unit, not just as individual words.

All languages have idiomatic expressions. However, phrasal verbs are specific to Germanic languages such as English [11]. Even the very concept of phrasal verbs may puzzle L2 readers if their first language is not Germanic. They may not realise these words are a phrasal unit and try to interpret the literal meaning of the individual words.

Phrasal expressions:

- are arbitrary. 'Give up' could just as sensibly be 'give down'—but it is not. 'Burn up' and 'burn down' are sometimes interchangeable—but not always. We say 'depend on' but not 'depend to' or 'depend by'—for no discernible reason.
- are not usually transparent or guessable. For example, we cannot guess the meaning of 'give up' by knowing the meaning of each of those words. 'Give' means to provide something. 'Up' indicates a higher place. But 'give up' does not mean to provide something to a higher place.
- have many meanings. For example, 'bring up' can mean 'nurture' ('bring up children'), 'vomit' ('bring up your food') or 'mention' ('bring up in conversation') [11].

The English language is rife with idiomatic expressions. In addition, each country, each region, each community has its own set of idioms. No wonder even highly advanced learners of English find it difficult to learn these [11]. Even if they can guess at the meaning, it slows down the flow of reading and leaves the reader with an unclear sense of the text.

Multiple Meanings

L2 readers have more difficulty with words that have several meanings [12]. As they are reading, they tend to understand a word's most common meaning, even if it doesn't make sense in the context. For example, L2 readers are more likely to interpret the word 'state' as a political entity. They are less likely to know the scientific sense of the word, 'a condition of matter'. Also, they are more likely to take the word 'since' to mean 'from a point in time' rather than 'because'. If one of the meanings of a word is idiomatic, it can cause even more confusion.

Difficult Pronunciation or Spelling

Words that are difficult to say or spell are more difficult to remember. Many readers sound out the words in their heads as they read silently. If they cannot do this easily, they may find it difficult to remember the words. A word that contains sounds that do not exist in the L2 reader's first language will be even more problematic. If you have ever read foreign novels in translation, you may know the difficulty of remembering characters' names. Long and seemingly unpronounceable names are harder to remember.

Spelling in English is notoriously difficult. There is not always a consistent relationship between the alphabet and the sounds. There are, of course, some rules —but also plenty of exceptions. For example, the spelling of 'fundamental' correlates well to its pronunciation. But the spelling of 'psychology' does not. Words that do not follow the rules are not easily sounded out. Thus, they may be more difficult for L2 readers to remember.

Words that appear similar in spelling may also cause confusion. For example, the words 'through', 'though' and 'thought' have different meanings and pronunciations. But they share most of the same letters.

Lack of Equivalent Concepts

If L2 readers do not have a certain word for a concept in their language, it may be harder to learn the word in English. Conversely, L2 readers may have many words in their first language for which English has none or only one. Both situations may make learning a word more difficult.

Lack of Similarity

L2 readers are more likely to know English words that are the same or similar to words in their own language [13]. For example, French, Italian or Spanish readers can more easily grasp Latin- or even Greek-based English words. It is easier to recognise words like 'physical', 'transmit', 'analysis' or 'ratio'. Chinese or Arabic languages have fewer words in common with English. However, many languages have 'borrowed' from English, especially modern science and technology words. For example, Arabic speakers use the terms 'malaria' and 'radar'.

4.3.4 Grammar Patterns

Grammar is not as big a factor as vocabulary in determining how well L2 readers can comprehend text. But L2 readers have to understand how words work together to make meaning [14].

There is little solid research on the effects of grammar on L2 reading comprehension. But extensive experience with L2 readers indicates that problem areas are modal verbs, complex noun phrases, and complex sentence patterns.

4.3.4.1   Modal Verbs

I would if I could but I can't, so I shan't.

The main modal verbs in English are listed in Table 4.2. Modal verbs have different functions—meanings, nuances and uses—depending on the structure of the sentence and the context in which they are used. The ticks in the table show the possible functions of each modal, but the list of functions is not exhaustive.

Table 4.2 Modal verbs and their functions

Function/ Modals	Suggestion	Possibility	Ability	Necessity	Permission	Obligation/ prohibition
Can	✓	✓	✓		✓	
Could	✓	✓	✓		✓	
May	✓	✓			✓	
Might	✓	✓			✓	
Will		✓				
Would		✓				
Should	✓			✓		✓
Must				✓		✓

Modals cause problems for many reasons, as Examples 4.5–4.8 highlight.

Different Modals, Same Meaning Several modals can express the same idea.

Example 4.5

> A successful scientific model <u>can</u> be made more sophisticated.
> A successful scientific model <u>could</u> be made more sophisticated.
> A successful scientific model <u>may</u> be made more sophisticated.

One Modal, Many Meanings One modal can express many different ideas. As shown in Table 4.2, 'can' has at least four uses.

Example 4.6

> *Possibility:* Inaccurate measurements <u>can</u> have disastrous results.
>
> *Permission:* You <u>can</u> complete the assignment tomorrow.

Past, Present or Future? A single modal can refer to the past, present or future, depending on its meaning and context.

Example 4.7

> *Past:* By the late 17th century geologists <u>could</u> accept that the earth had a much longer history than had previously been thought.
>
> *Present:* The current theory of plate tectonics <u>could</u> explain a number of geologic phenomena.
>
> *Future:* Continued fossil matching studies <u>could</u> still lead to refinements in the theory of plate tectonics.

Negative, Not Always Opposite The negative form of a modal does not always mean the opposite. In Example 4.8, 'must' indicates necessity or obligation, but 'must not' can only refer to prohibition, not lack of necessity. To indicate lack of necessity, we need to write 'don't have to' (i.e. not necessary).

Example 4.8

Necessity: You <u>must</u> finish it.

Prohibition: You <u>must not</u> finish it.

Not a necessity: You <u>don't have to</u> finish it.

4.3.4.2   Complex Noun Phrases

A noun phrase is a noun (a person, place, thing or idea) and any other words that describe or modify it. In English, several words can describe a noun. This often results in long noun phrases that contain densely packed information. Formal academic texts are typically littered with such noun phrases. Example 4.9 shows a sentence that contains two complex noun phrases.

Example 4.9

<u>Each of the estimated 100 trillion cells in the adult human body</u> contains a <u>network of long fibres making up the cell's structural framework.</u>

The main noun of the first phrase is 'each' (referring to cells). The main noun of the second phrase is 'network'. But much more information about the cell and the network is packed into these noun phrases. This forces the reader to work harder to unravel all that information.

L2 readers find complex noun phrases difficult because:

- the main noun is buried among other words and phrases, forcing readers to 'unpack' the information to understand the meaning;
- nouns often serve as adjectives, making it doubly difficult for readers to identify the main noun;
- the words can connect to each other in different ways, creating ambiguity;
- the phrases can cause the main noun to be far from its corresponding verb, causing readers to lose track of who is doing what in the sentence. In the example above, 11 words stand between the main noun 'Each' and the verb 'contains'. Readers must hold in their minds a greater amount of information as they move from the subject to the verb.

4.3.4.3   Complex Sentence Patterns

Sentences with many phrases and clauses are generally harder for L2 readers. In such sentences, it is often difficult to determine who or what is doing what. With so much going on in the sentence, it is hard to match the various subjects to their verbs.

Even with just two clauses, the sentence in Example 4.10 can cause confusion.

Example 4.10

Carbon dioxide-producing energy technologies the professor talked about have the potential to reduce global greenhouse gas emissions.

The two clauses are:

- Carbon dioxide-producing energy technologies have the potential to reduce global greenhouse gas emissions;
- (that) the professor talked about.

The sentence is perfectly grammatical. But one clause is embedded in the middle of the other. In addition, the linking word 'that' is dropped. Without this grammatical clue, it is more difficult for L2 readers to recognise the second clause in the sentence.

Another problem is that the second clause is embedded in the first clause. This causes the subjects of both clauses to appear sequentially ('technologies' and 'professor'). The verbs of each clause also appear together later in the sentence ('talked about' and 'have'). This makes it harder to match up the subjects with their corresponding verbs.

Example 4.11 shows a sentence that contains four clauses.

Example 4.11

Initially, the cracking process was done using thermal methods, but today, the catalytic process has mostly replaced thermal cracking because it produces more gasoline having a higher octane and reduced amounts of heavy fuel oils and light gases.

The four clauses are:

- Initially the cracking process was done using thermal methods;
- but today the catalytic process has mostly replaced thermal cracking;
- because it produces more gasoline and reduced amounts of heavy fuel oils and light gases;
- (that has) having a higher octane.

The fourth clause is confusing for two reasons. It falls in the middle of the third clause, and it is in short form. Without the signal word 'that', it is harder to identify it as a clause that describes the preceding word. L2 readers tend to learn the formal, longer forms of sentences first and the shorter versions later. Thus, such reduced clauses are often harder to understand.

4.3.5   Text Flow

Sentences in a text are organised in a particular way to make the information flow. They are also meaningfully related to each other to show the connection between ideas. The techniques used to make a text flow vary from language to language. These differences can affect the L2 readers' ability to make sense of a text, especially in academic contexts.

Common L2 reading problems stem from an inability to recognise that some words refer to others. Words like 'she', 'he', 'it', 'one' and 'this', 'that' 'here', 'there', et cetera are used to refer to meaningful words or phrases in other parts of the text. Sometimes it is fairly obvious what these words refer to. But sometimes, readers must do the work of making the connections.

Also, making those connections can depend on knowing the other words in the text. Thus, L2 readers with a limited vocabulary will find this even harder to do.

Examples 4.12–4.15 show how shorter words can refer to other words, and how this can slow down the reading process.

In Example 4.12, 'they' and 'them' replace a plural noun.

Example 4.12

The theoretical models emerge from various experimental studies of the system.
Mostly unproven, they tend to reflect the major schools of thought at the time.
A historical timeline shows them in neat progressive development, but the facts
emerged in a more random fashion.

There are two plural nouns in the previous sentence: 'theoretical models' and 'experimental studies'. But the intended reference is to 'theoretical models'. The reader must infer this using the main clue that theoretical models—and not studies —can be proven or not proven.

In Example 4.13, the word 'so' refers to an entire phrase.

Example 4.13

The English scientist, John Dalton, first proposed the atomic theory to account for
the experimental observations reported at the time. By doing so, he was able to
carry out the earliest predictive studies of new compounds.

Readers must infer that 'so' refers to the proposal of the atomic theory. This is a very specialised use of the word 'so'. This word has other functions and meanings that L2 readers are more likely to be familiar with.

In Example 4.14, the words 'part' and 'remainder' refer back to the subject of the sentence.

Example 4.14

Radiant energy from the Sun reaches the Earth as electromagnetic waves. Part of it is absorbed. Another part is reflected. The remainder goes past the Earth.

In Example 4.14, 'Part', 'Another part' and 'The remainder' refer to 'radiant energy from the Sun'. This makes for less repetitive writing. However, it can be harder for L2 readers to recognise that these three phrases all refer to a part of a whole.

Example 4.15 shows how different words are used in different sentences to refer to the same thing.

Example 4.15

Minerals are all solid at the earth's surface temperature. This matter has orderly crystal structures.

The word 'minerals' is referred to in the subsequent sentence as 'matter', a broader term. This makes the text less repetitive. But for an L2 reader, it can make it more difficult to connect one word to the other. They may not make the connection immediately or they may not know the meaning of the alternate terms.

4.3.6   Text Organisation

Text organisation refers to the preferred ways in which texts:

- present and develop main ideas;
- make persuasive arguments;
- present evidence;
- use repetition and paraphrasing;
- organise new information;
- expect the reader to critically interpret or infer material [15, 16].

Ways of organising texts vary from language to language. The purpose and organisation of a text in English may not match L2 readers' expectations. This can have a huge impact on the L2 readers' ability to handle an academic reading text. For example, certain types of information may not be where they expect it.

This kind of problem is invisible. Readers can easily recognise when individual words are unfamiliar, whereas they are not always aware that their culturally embedded expectations are at odds with the discourse of a English text.

For example, one common problem area for L2 readers is recognising definitions in texts. There are several ways of indicating the meaning of new or unfamiliar terms in English. But L2 readers often overlook them. Instead, they go straight to their dictionaries. This is not only a waste of time, but it can also lead to poor comprehension. In addition, the dictionary definitions of words do not always match the scientific meanings. The in-text definitions are more accurate and specific.

4.4 Making Texts Readable

As writers, we must keep in mind the perspective of L2 readers. By addressing some of their potential problems, we can reduce the burden on their bottom-up reading processing. This frees up readers' working memory so that they can focus more on the meaning and less on decoding. In this way, we can make texts more readable.

Making texts readable does not mean oversimplifying or dumbing down the content. It means using language in such a way that content becomes clearer.

References

1. C.E. Nuttall, *Teaching Reading Skills in a Foreign Language*, 2nd edn. Macmillan books for teachers (Macmillan, 2005)
2. W.P. Grabe, F.L. Stoller, Comparing L1 and L2 reading, in *Teaching and Researching: Reading*, ed. by C.N. Candin, D.R. Hall (Routledge, London, New York, 2013), p. 50
3. W.P. Grabe, Reading in different languages, in *Reading in a Second Language—Moving from Theory to Practice*, ed. by M.H. Long, J.C. Richards (Cambridge University Press, Cambridge, 2009), p. 123
4. W.P. Grabe, Reading in different languages, in *Reading in a Second Language—Moving from Theory to Practice*, ed. by M.H. Long, J.C. Richards (Cambridge University Press, Cambridge, 2009), p. 112
5. W.P. Grabe, Reading in different languages, in *Reading in a Second Language—Moving from Theory to Practice*, ed. by M.H. Long, J.C. Richards (Cambridge University Press, Cambridge, 2009), pp. 114–118
6. J.S. Barrot, Revisiting the role of linguistic complexity in ESL reading comprehension. 3L: Southeast Asian J. Engl. Lang. Stud. **19**(1), 5–18 (2013)
7. K.M. Hummel, *Introducing Second Language Acquisition* (Wiley, Chichester, 2014), p. 148
8. N. Schmitt, X. Jiang, W. Grabe, The percentage of words known in a text and reading comprehension. Modern Lang. J. **95**(1), 26–43 (2011)
9. I.S.P. Nation, How large a vocabulary is needed for reading and listening. Can. Modern Lang. Rev. **63**(1), 59–82 (2006)

10. J. Milton, The development of vocabulary breadth across the CEFR levels, in *Communicative Proficiency and Linguistic Development: Intersections Between SLA and Language Testing Research*, ed. by I. Bartning, M. Martin, I. Vecder (European Second Language Association, 2010), pp. 211–232

11. A. Siyanova, N. Schmitt, Native and nonnative use of multi-word versus one-word verbs. IRAL Int. Rev. Appl. Linguist. Lang. Teach. **45**, 119–139 (2007)

12. B. Laufer, What's in a word that makes it hard or easy: some intralexical factors that affect the learning of words, in *Vocabulary: Description, Acquisition and Pedagogy*, ed. by N. Schmitt, M. McCarthy (Cambridge University Press, Cambridge, 1997), pp. 140–155

13. Education First, *EF English Proficiency Index* (2016) (cited 2018), www.ef.com/epi

14. M.J. Heilman et al., Combining lexical and grammatical features to improve readability measures for first and second language texts, in *Proceedings of NAACL HLT* (2007), pp. 460–467

15. W.P. Grabe, L1 and L2 reading relationships, in *Reading in a Second Language—Moving from Theory to Practice*, ed. by M.H. Long, J.C. Richards (Cambridge University Press, Cambridge, 2009), p. 139

16. W.P. Grabe, F.L. Stoller, Comparing L1 and L2 reading, in *Teaching and Researching: Reading*, ed. by C.N. Candin, D.R. Hall (Routledge, London, New York, 2013), p. 43

Chapter 5
What's in a Word?

Contents

Abstract What's in a word? Plenty. To read fluently and understand adequately, L2 readers need to know 98% of the words in a text. Thus, a good starting point for making texts readable is to choose words carefully. This chapter discusses a number of useful guidelines for increasing your reader's word recognition.

Keywords Simplified vocabulary · Readable texts · Readable science · Reading comprehension

5.1 Use High-Frequency Vocabulary

Except for essential technical words, aim to use the most frequent words in English. L2 readers are more likely to know them because higher frequency words are more likely to be acquired first. In addition, such words have generally been the focus of English language programmes since the 1990s [1]. Therefore, L2 readers process high-frequency words in a text more quickly and understand them better than low-frequency words.

A reasonable goal would be to use fewer than 4000 word families to cover 98% of your text. A word family consists of:

- a root word—example: 'inform';
- its inflections—example: 'informs', 'informed', 'informing';
- its derivations—example: 'information', 'informative'.

Frequency word lists are freely available online. But there is no need to consult these long lists as you write. Online vocabulary profilers are also available to scan texts and categorise each word according to its frequency in English. This makes it easy to identify and replace low-frequency words with high-frequency ones. Chapter 10 shows how to use this tool.

When choosing words, it is important to balance frequency level with other factors. For example, higher frequency words tend to have several meanings. And such words can reduce the clarity of a text rather than increase it [2].

5.2   Use the Necessary Technical Terms

You do not need to sacrifice key scientific vocabulary in order to increase the readability of texts. Use the technical terms that you need in your text.

All science and technology fields need specialised words to express ideas and concepts efficiently. And L2 readers need these words to function in their chosen fields. In addition, any attempt at simplifying technical vocabulary can result in loss of clarity—and can even lead to the ridiculous.

The paragraph in Example 5.1 shows how confusing it can get when technical words are avoided. The text promises to get worse as it goes on to discuss other layers of the atmosphere without benefit of three key technical words—'troposphere', 'stratosphere' and 'vapour'. A revised version with these words included brings much clarity. A suitable diagram would also be helpful.

Example 5.1

Before: The lowest layer of Earth's atmosphere contains about 80% of the atmosphere's mass and 99% of the total mass of clouds and water as a gas. The lowest layer of Earth's atmosphere ranges from 7 to 20 km wide. The conditions that occur in the lowest part of the lowest layer of Earth's atmosphere are called weather. The lowest part of the lowest layer of Earth's atmosphere is the layer that forms the boundary with Earth. On top of the lowest layer of Earth's atmosphere is the boundary between the lowest layer of Earth's atmosphere and the second main layer of Earth's atmosphere …

After: The **troposphere** contains about 80% of the atmosphere's mass and 99% of the total mass of **vapour** and clouds. The **troposphere** ranges from 7 to 20 km wide. The conditions that occur in the lowest part of the **troposphere** are called weather. The lowest part of the **troposphere** is the layer that forms the boundary with Earth. On top of the **troposphere** is the **tropopause**, which is the boundary between the **troposphere** and the **stratosphere** …

Technical words are not always necessary. Decide which technical words are necessary in your field—and which are not. If they are necessary, define them. If they are not, leave them out. Avoid coining new expressions or words, unless they add clarity or precision to the text and the subject matter.

5.3   Define New Words Consistently

There are several techniques for defining words within a text. But many L2 readers do not notice or recognise these techniques and therefore, often overlook the definitions. They end up wasting their time looking up words that are already defined for them on the page.

To make it easier for the L2 reader, choose one obvious way of defining technical terms—and stick with it. Example 5.2 shows some techniques for clearly defining terms.

- put the definition in a separate box, but within the normal flow of the text;
- use a straightforward style of: 'A **fusswudget** is a hammer with five dots on it';
- highlight the term using boldface or a special colour that is not used for any other purpose in the text;
- highlight that term each time it is used throughout the text;
- consider including the pronunciation in parentheses or, in a digital version, adding a live link to a sound file;
- include a glossary of all highlighted terms at the end of the text or, in a digital version, as a live link.

Example 5.2

All organisms contain a common set of biological molecules. They are composed of cells, and can maintain **homeostasis**.

Homeostasis (home-ee-oh-STAY-sis) is the ability to maintain internal conditions despite constant change in the external environment.

In addition, populations of living organisms can evolve …

Avoid using italics, as this font style is more difficult to read. Also avoid using footnotes. They tend to break the flow of reading, forcing readers to come back and find their place in the text after chasing down the definition.

Since textbooks are not always used linearly, you may want to consider repeating definitions in each section or chapter as needed.

5.4   Avoid Using Technical Terms in a Non-technical Way

Reserve subject-specific technical words for their primary technical meaning. If a common word is used in a specific way in your field, avoid using that word for its more common meaning.

For example, if you use the word 'solution' to refer to a molecular mixture of two or more substances, do not use it to refer to the process of solving a problem.

Choose another word to express the common meaning, such as 'answer', 'key' or 'explanation'.

Another example is the word 'force'. In almost any science or technology text, the word 'force' should refer only to pressure per square metre. It should not be used in the sense of 'strength', 'power' or 'push hard'.

5.5 Use Words in Their Literal Sense

Avoid idiomatic expressions and figurative language as much as possible.

The more colourful idiomatic expressions are easy to identify and avoid. Just a few examples are:

- hold your horses;
- slipped between the cracks;
- kick the bucket;
- cool as a cucumber;
- a piece of cake.

These expressions are usually clichés, or they are specific to the speech of certain geographic areas. Therefore, they generally have no place in scientific writing anyway.

However, some figurative language is not so easy to identify and avoid because it has become part of everyday language. For example, the word 'sharp' has come to mean 'intelligent', and 'cutting edge' to mean 'newest'. Beware these kinds of expressions. They can cause confusion, especially for readers who do not know all the meanings of a word. L2 readers most often know only the literal meanings.

In addition, if L2 readers need to look up a word in a bilingual dictionary, the entry will probably list the literal meaning first. Thus, it is better to stick to the literal uses of words as much as possible.

5.6 Choose Words with the Clearest Meanings

Choose the words that have the most specific definitions. For example, something that is not easy is often referred to as 'hard'. This word may seem a better choice than 'difficult', which is longer and more formal. But the main meaning of 'hard' is 'not soft' or 'firmly formed'. So although 'hard' and 'difficult' are both high-frequency words, 'difficult' is a better choice. It is more specific and less ambiguous. Reserve the word 'hard' for physical properties.

Instinctively, we might think shorter words are easier. This is more likely to be true for L1 readers, but not necessarily so for L2 readers, as the example above shows. Short, one-syllable words often have so many meanings, many of them figurative.

Also, choose the form of the word that has the most specific meaning. Sometimes words are specific in one form but not in another. For example, the word 'address' is quite specific as a noun. But when it is used as a verb, it becomes vague. When we address a problem or an issue, are we talking about it? Solving it? Explaining it? Or merely identifying it?

5.7   Avoid Phrasal Verbs

A phrasal verb consists of a verb and a preposition and/or adverb. It has a specific meaning that is usually different from its individual words. For example, 'run out of' has nothing to do with the activity of running.

There are thousands of such expressions in English. And they can be confusing for L2 readers. Where possible and sensib.e, use single verbs instead. Table 5.1 shows a few examples of phrasal verbs that can be replaced with a single word. The words in both columns fall within the 3,000 most common words in English.

Table 5.1 Single word alternatives to phrasal verbs

Instead of	Try using
add up	calculate; total
call for	need; require
carry on	continue
carry out	do; complete; accomplish
cut back	remove; reduce
cut out	remove
find out	discover; find
go on	continue
hold up	delay
leave out	not use; exclude
look into	research; probe
make clear	clarify; explain
make sure (of)	ensure; prove
make up	make; prepare; invent; create
pass on	transmit
point out	explain; focus
set up	arrange
use up	use all of; exhaust; finish

A phrasal verb can sometimes be stripped down to the main verb without changing the meaning significantly, as shown in Example 5.3.

Example 5.3

Phrasal verb: Our immune system is designed to <u>fight off</u> infections.

Single verb: Our immune system is designed to <u>fight</u> infections.

Phrasal verb: <u>Make up</u> a concentrated solution of acid.

Single verb: <u>Make</u> a concentrated solution of acid.

Sometimes, phrasal verbs are perfectly transparent and there is no need to reduce or replace them, as shown in Example 5.4.

Example 5.4

It was difficult to <u>lift up</u> the heavy container to the required height.

<u>Spread out</u> the crystals on filter paper and allow them to dry at room temperature.

If you <u>take away</u> the heat source, crystallisation will begin.

Sometimes, choosing a single word over its phrasal equivalent may seem counter-intuitive if the single word is longer, more formal and lower frequency. But longer words, especially if they are more specific, are not necessarily more difficult for L2 readers. In fact, for some readers, especially Europeans, the longer Latin-based words are easier [3].

For example, the words 'pass' and 'on' in the phrasal verb 'pass on' are both short and frequent, falling within the 1000 most frequent words. The word 'transmit' is longer, more formal and a little less frequent, falling within the 3000 most frequent words. But it is more specific. In addition, because it derives from the Latin word 'transmittere', it is more—not less—familiar to many L2 readers, especially Europeans.

If you cannot avoid using phrasal verbs, try to use high-frequency ones. These are the ones that L2 readers are more likely to know. High-frequency phrasal verbs are listed in Table 5.2 [4, 5]. This list has been edited to include only those phrasal verbs that are likely to appear in academic and science texts. Note: these verbs can be used in any form as needed (singular/plural, past/present, active/passive, etc.).

If you use a phrasal verb, use its most common meaning. Some phrasal verbs have several meanings [6]. The examples in Table 5.2 show the most common meaning of each phrasal verb. L2 readers are more likely to know these.

In addition, if you use a phrasal verb, keep its parts together, as shown in Example 5.5. Do not separate them.

Example 5.5

Before: You need to <u>set</u> the experiment <u>up</u>.

After: You need to <u>set up</u> the experiment.

The frequency of phrasal verbs is a relatively new area of research, so they have not yet been incorporated into vocabulary profilers. However, development is underway to do so [1].

Table 5.2 Most common phrasal verbs

Phrasal verb	Example
account for	Experimental design must account for external conditions.
aim to	Analytical procedures aim to provide useful data.
be (is/are) based on	The report is based on the data.
be (is/are) bound to	Carefully controlled studies are bound to be more accurate.
be (is/are) concerned with	The scientific method is concerned with finding information.
be (is/are) due to	Some believe it is due to global warming.
be (is/are) expected to	The results are expected to provide the final answers.
be (is/are) followed by	Product isolation is followed by purification.
be (is/are) found to	Food contamination was found to be the cause.
be (is/are) known to	Several chemical agents are known to cause cancers.
be (is/are) likely to	Rigorous testing is likely to provide a solution.
be (is/are) meant to	Heating was meant to improve consistency.
be (is/are) subject to	Animal behaviour is subject to environmental factors.
call for	Global problems call for international cooperation.
carry out	This is not a difficult procedure to carry out.
come back	We will come back to this topic later.
come out	The next article will come out next Friday.
consist of	What does this material consist of?
deal with	The researcher has several key issues to deal with.
end up	Chemical waste can end up in groundwater supplies.
focus on	Physical geologist focus on the materials of the Earth.
follow up	Scientists follow up observations to check reproducibility.
go back	Historical geology allows you to go back to the Earth's origins.
go through	Pharmaceutical products must go through rigorous testing.
lead to	Smoking can lead to heart disease.
make sure	Stress testing makes sure that the product is safe.
make up (comprise)	The chemical elements make up all matter.
manage to	Bears manage to survive the hibernation period.
point out	Survey results point out current trends.
rely on	Scientists must be able to rely on their instruments.
seek to	Einstein sought to explain relativity.
set up	You need to set up the experiment.
take into account	Population control must take into account many factors.
take part in	Volunteers are paid to take part in clinical trials.
take place	The reaction will take place spontaneously.
tend to	Multiple repetitions tend to produce greater accuracy.
result in	Trace contaminants can result in toxic reactions.
work out	Try to work out the problem.

5.8   Use Words Consistently

Use the same words for the same thing throughout your text. If there is more than one word to express the same idea, choose one and stick with it. For example, 'finish' and 'complete' can usually be used interchangeably. So can 'but' and 'however'. It is better to choose one of the pair and use it consistently.

For L2 readers, consistency is much more important than variety. Variety can be unnecessarily confusing. For example, readers might erroneously conclude that the different words mean different things, when in fact they refer to the same thing.

Maintaining consistency in the use of words also helps reduce vocabulary diversity, which is the range of words used in a text. It is a common measure of readability expressed as a ratio of the number of *different* words to the total number of words in the text. The more words that are repeated in a text, the lower the vocabulary diversity. The lower the vocabulary diversity of a text, the easier it is to read [7].

5.9   Use Hyphens Liberally

Consider hyphenating more often than you normally would. This helps L2 readers recognise words more easily. It also makes it easier for them to find the root word in a dictionary or glossary if necessary [8].

Examples 5.6–5.8 show different situations in which it is useful to hyphenate between prefixes and their stems.

- Clarify low-frequency or technical words.

Example 5.6

```
pretreat → pre-treat
rehybridization → re-hybridization
```

- Distinguish between similar words.

Example 5.7

```
re-cover (give something a new cover)
recover (get something back)
```

- Separate vowels where the prefix and stem are joined.

Example 5.8

deice → de-ice
cooperate → co-operate
antioxidant → anti-oxidant
reintegrate → re-integrate

Prefixes are commonly taught in English language courses, so L2 readers will most likely recognise common prefixes, such as:

– anti, de, dis, un, non, in (to indicate opposite or lack of)
– co, inter, mis, mid, re, over, pre, pro. super, semi, sub, trans, under.

Examples 5.9–5.11 show how to use hyphens between closely related words.

• Connect letters or numbers to a stem word.

Example 5.9

L shaped → L-shaped
3 prong connector → 3-prong connector

• Connect multiple-word adjectives.

Example 5.10

high volume production → high-volume production
high pressure chamber → high-pressure chamber
air conditioned compartment → air-conditioned compartment
trial and error method → trial-and-error method
three to one ratio → three-to-one ratio
carbon neutral → carbon-neutral

• Connect parts of a compound verb.

Example 5.11

fast forward → fast-forward

cold roll → cold-roll

vacuum pack → vacuum-pack

short circuit → short-circuit

Current publishing practice avoids hyphenating words. This may render a cleaner look, but sometimes, it makes it difficult for L2 readers to recognise the stem word or to recognise word clusters. If you are working with publishers, you may need to convince them of the value of maintaining the hyphens.

References

1. T. Cobb, FREQUENCY 2.0: incorporating homoforms and multiword units in pedagogical frequency lists, in *L2 Vocabulary Acquisition, Knowledge and Use: New Perspectives on Assessment and Corpus Analysis*, ed. by C. Bardell, C. Lindquist, B. Laufer (EUROSLA, 2013), p. 94
2. S.A. Crossley et al., A linguistic analysis of simplified and authentic texts. Modern Lang. J. **91**, 15–30 (2007)
3. Education First, *EF English Proficiency Index* (2016) (cited 2018), www.ef.com/epi
4. R. Martinez, N. Schmitt, A phrasal expressions list. Appl. Linguist. **33**(3), 299–320 (2012)
5. M. Garnier, N. Schmitt, The PHaVE list: a pedagogical list of phrasal verbs and their most frequent meaning senses. Lang. Teach. Res. **19**(6), 645–666 (2015)
6. M. Garnier, N. Schmitt, Picking up polysemous phrasal verbs: how many do learners know and what facilitates this knowledge? System **59**, 29–44 (2016)
7. P.M. McCarthy, S. Jarvis, MTLD, vocd-D, and HD-D: a validation study of sophisticated approaches to lexical diversity assessment. Behav. Res. Methods **42**(2), 381–392 (2010)
8. E.H. Weiss, Reducing burdens, in *The Elements of International English Style* (M. E. Sharpe, New York, London, 2005), p. 73

Chapter 6
Get a Grip on Grammar

Contents

Abstract Science texts need to be precise. This precision usually requires detailed descriptions of objects, forces, organisms, and methodology as well as extensive explanations of complex concepts. It is no wonder then that science texts are often densely packed with complicated sentences that try to express many ideas. L2 readers struggle with complicated grammatical structures and tortuous sentences. This chapter provides guidelines for keeping sentence structures simple and readable while conveying complex information.

Keywords Sentence structure · Simplified grammar · Simple grammar · Writing clarity

6.1 Keep Noun Phrases Short

Avoid long noun phrases by breaking up the information into shorter sentences. This might result in a less concise text, but it will be clearer. Breaking up noun phrases also makes the text less dense and therefore easier for the readers to absorb.

The underlined portion of the sentence in Example 6.1 is a noun phrase. The whole phrase is the subject of the sentence. However, the main nouns, 'research and development', are buried in the phrase.

We can reveal the main nouns by breaking up the sentence. The revision may seem less efficient and more repetitive. But it is much clearer for L2 readers.

Example 6.1

Before: The extensive research and development required to provide the appropriate safety and effectiveness of new drug products is complicated, costly and time consuming.

After: Extensive research and development is required to provide safe and effective drugs. This research and development is a complicated, costly and time-consuming process.

In Example 6.2, there are two long noun phrases. The first one is the subject of the sentence and the second is the object. As in Example 6.1, we can reveal the main nouns by breaking up the sentence.

Example 6.2

Before: Each of the three main reaction types in organic chemistry has several possible condition-dependent mechanistic pathways.

After: There are three main reaction types in organic chemistry. Each reaction type has several possible mechanistic pathways. These pathways depend on the conditions of the reaction.

Sometimes, long noun phrases involve names of people and places. We can clarify such noun phrases by removing the names if they are not essential.

Example 6.3

Before: Anders Celsius (1701–1744), a Swedish astronomer, created the Celsius scale in 1742.

After: A Swedish astronomer created the Celsius scale in 1742.

Keeping noun phrases short also helps keep the subject close to its verb, which brings us to the next guideline.

6.2 Put the Subject and its Verb Close Together

Once the readers have identified the subject of the sentence, they look for the verb that describes what the subject is doing. If the verb is too far from the noun, the reader may lose track of who or what is doing the action. This interrupts reading flow.

In Example 6.4, 30 words and two clauses lie between the main subject (underlined) and its verb (double-underlined). We can solve this problem by breaking the information into a series of shorter sentences.

Example 6.4

Before: All changes in the world around us, from the simple change of liquid water to water vapour when it is boiled to the complex changes in our bodies when they fight viruses, involve atoms and molecules (34 words, 1 sentence).

After: All changes in the world around us involve atoms and molecules. This is true of simple or complex changes. For example, a simple change is liquid water to water vapour when it is boiled. A complex change is when our bodies fight viruses (41 words, 4 sentences).

The revision contains more sentences and more words, but it is clearer without sacrificing content.

6.3   Choose the Simplest Verb Forms

Avoid using more complicated verb tenses when a simple one does the job. Use the following three tenses as much as possible. They are the simplest because they each contain a single word and because they are the ones that L2 readers usually learn first:

- simple present;
- simple past;
- command (for giving instructions).

Example 6.5 shows the different verb tenses (double-underlined).

Example 6.5

Simple present: Each time we discover a new species, we name it.

Simple past: When he examined a particular petri dish, Alexander Fleming observed an area around a mould that had no bacteria.

Command: Check the safety precautions before you try this procedure.

Luckily, these tenses also happen to be the most common forms used in scientific texts. However, sometimes, the future and the present perfect tenses are overused in situations that do not require them.

Example 6.6

Before: The compound with the greater molar mass <u>will have</u> the higher
boiling point.

After: The compound with the greater molar mass <u>has</u> the higher boiling point.

Example 6.7

Before: The manufacturer <u>has supplied</u> the safety procedures.

After: The manufacturer <u>supplied</u> the safety procedures.

Use other verb tenses only if they are essential to the meaning, as shown in
Example 6.8.

Example 6.8

Future tense: Astronomers expect that there <u>will be</u> improvements in the
Big Bang theory.

Present perfect tense: Since 1927, astronomers <u>have focused</u> their investigation
on the Big Bang model.

Present continuous: Hubble's law states that the more distant galaxies <u>are
receding</u> faster from us.

Avoid wordy verb constructions if there is another simpler way to express the
same ideas.

Example 6.9

Before: Long-period comets, such as Hale–Bopp, <u>are thought to originate</u>
in the Oort cloud.

After: Astronomers <u>believe</u> that long-period comets, such as Hale–Bopp, <u>originate</u>
in the Oort cloud.

Example 6.10

Before: Body Mass Index (BMI) <u>has come to be used</u> for preliminary diagnosis of obesity.

After: Body Mass Index (BMI) <u>is now used</u> for preliminary diagnosis of obesity.

Example 6.11

Before: Minerals <u>are able to transmit</u> light to various extents.

After: Minerals <u>transmit</u> light to various extents.

6.4   Use Passive Verbs Sparingly

Use active verbs unless the passive is necessary. Sentences with active verbs are more likely to be direct and clear.[1] Sentences with passive verbs can make a text more vague, ambiguous and even pompous.

Traditionally, academic and scientific writing guidelines encouraged the use of passive verbs. They can make texts seem more formal and objective. Passive verbs also allow writers to avoid using 'I' and 'we'.

However, modern scientific writers and editors recognise that passive verbs are overused. Even journals such as *Science* and *Nature* now encourage writers to use active verbs and no longer discourage the use of 'I' and 'we' in sentences [1]. The difference in styles is shown in Example 6.12.

Example 6.12

Passive: The samples <u>were cooled</u> on ice and then <u>frozen</u> until they <u>were needed</u>.

Active: We <u>cooled</u> the samples on ice and then <u>froze</u> them until we <u>needed</u> them.

However, passive verbs are sometimes necessary. Use passive verbs in the following two key situations:

---

[1]In a sentence with an active verb, the subject *does* the action, and the 'doer' appears at the beginning of the sentence. In a sentence with a passive verb, the subject *receives* the benefit of the action, and the 'receiver' appears at the beginning of the sentence. Sometimes, the person or thing that is responsible for the action is not mentioned at all if it is not important.

The road was crossed by the chicken.

- when the person or thing doing the action in the sentence is obvious, unknown or irrelevant. In Example 6.13, it would be difficult—or at best awkward—to make an active sentence using the verb 'segment'.

Example 6.13

The surface of the earth <u>is segmented</u> into about 20 plates.

Who segmented the surface of the earth? Does it matter? Probably not. In Example 6.14, the writer does not know or care who did the action.

Example 6.14

A large concentrated solar power plant <u>was built</u> for Masdar City in the desert city of Abu Dhabi.

The focus is on the solar power plant, not on who built it. However, if this sentence appears in a trade magazine for solar power engineers, the active verb is probably better. The readers are more likely to be interested in the name of the company that built the plant.

- when the receiver of the action is more important than the person or thing that did it. In this case, the passive voice allows writers to focus on *what* was found rather than *who* found it. In Example 6.15, it is not important who made the discovery unless it is a text specifically about NASA. The discovery—methane in Mars' atmosphere—is more important.

Example 6.15

Passive: Methane <u>was discovered</u> in Mars' atmosphere.

Active: NASA scientists <u>discovered</u> methane in Mars' atmosphere.

6.5   Use Modals Consistently

Modal verbs—such as 'can', 'could' and 'may'—are important in scientific writing. They help qualify or modify information without loss of precision. They indicate degrees of possibility, certainty and doubt.

Most modal verbs serve more than one function. One modal may have several different meanings, nuances or uses. The function of a modal depends on the structure of the sentence and the context in which the modal is used. L1 readers understand the various functions instinctively. But L2 readers struggle with their complexities.

As much as possible, choose one modal for each function and use it consistently and exclusively for that function. In Example 6.16, three different modals are used unnecessarily.

Example 6.16

Before: A volcano <u>may</u> produce gases at any time during its life span. Ninety per cent of these gases <u>can</u> be steam, but carbon dioxide and hydrogen sulphide are also present. A volcano <u>could</u> also produce various amounts of lava.

After: A volcano <u>may</u> produce gases at any time during its life span. Ninety per cent of these gases <u>may</u> be steam, but carbon dioxide and hydrogen sulphide are also present. A volcano <u>may</u> also produce various amounts of lava.

Table 6.1 is a summary of suggestions for managing the potentially confusing array of modal verbs. The table is followed by further details, explanations and examples.

Some suggestions for managing the potentially confusing array of modal verbs:

Table 6.1 Summary of suggested modal use

Use	Rather than	To express only	But not to express
can	• could • might • be able to	• ability or possibility in present or future	• permission • polite request
cannot	• could not • might not • not be able to	• inability or impossibility in present or future	• permission • polite request
may/may not	might	uncertain possibility	• permission • polite request • wishes ('May the force be with you'.)
could/could not	was able to	past ability	• possibility • permission • polite request
will	be going to	• simple future • conditional sentences if needed	• what always happens in certain situations • habitual inclination to do something • certainty that something is true • capability (This tree will survive for 3 months.) • promise/offer/intention
will not		• negative simple future • conditional if needed	negative of above
would	(use sparingly)	• hypothetical, unreal or untrue situation • unreal conditional	• permission • polite request • habitual action in the past
would be able to	could	hypothetical ability or possibility	
should	ought to (use sparingly)	expected or recommended behaviour or situation	• deduction or high degree of certainty (The results should be ready by now.) • replacement of 'if' in conditional sentence (Should the volcano erupt, it would kill thousands of people.)
must	• should • have to • will need to • will require	• necessity • requirement	• reasonable expectation or logical conclusion (The white residue must be the result of calcium deposits from the dripping water.)
must not	should not	prohibition	negative of the above

- *Use 'can' to express ability or possibility in the present or future.* This function includes the power or capacity to do something as well as the potential for something to happen. 'Can' is shorter, simpler and more flexible than the alternatives 'could', 'might' or 'be able to'.

Example 6.17

With modern telescopes we <u>can identify</u> at least a billion galaxies.

Scientists classify objects by similar properties so that they <u>can study</u> their physical characteristics more easily.

- *Use 'cannot' to express inability or impossibility in the present or future.*

Example 6.18

You <u>cannot see</u> the Horsehead Nebula without a telescope.

Alpha particles <u>cannot pass</u> through clothing or skin.

- *Use 'may'/'may not' to express an uncertain possibility in the present or future.* 'May' is simpler and more flexible than the alternative 'might.' NOTE: When expressing uncertain possibilities, 'may not' is not the opposite of 'may'. Rather, the difference is in the perspective—optimistic or pessimistic.

Example 6.19

The sample size <u>may be large</u> enough to be statistically significant. (We hope that it is large enough.)

The sample size <u>may not be</u> large enough to be statistically significant. (We are doubtful that it will be large enough.)

To ensure clarity of the pessimistic stance, try rewriting the sentence as follows.

Example 6.20

It is possible that the sample size is not large enough to be statistically significant.

- *Use 'could'/'could not' to express an ability/inability in the past.*

Example 6.21

Where the soils were more basic, they could neutralize more acid rain.

By 2000, researchers at the CDC could detect TCDD at levels of 10^{-16} g.

Before 2000, researchers could not detect TCDD at such low levels.

- *Use 'will'/'will not' to express an action in the future and in conditional sentences where required.*

Example 6.22

Future: The environmental load will continue to recycle among air,
land and water for many decades in the future.

Conditional: If coral remains bleached for a long period of time, it will probably die.

In science, 'will' is often used to show what always happens in certain situations. But in most such instances, the present simple does just as well—and is simpler.

Example 6.23

Before: The oestrogen receptor will bind to a number of compounds.

After: The oestrogen receptor binds to a number of compounds.

Example 6.24

Before: Plants will grow more quickly when they are fertilised.

After: Plants grow more quickly when they are fertilised.

- *Use 'would' to express an action that is hypothetical, unreal or untrue, often in conditional sentences. But use it sparingly. Guideline 6.10 discusses the use of conditional sentences in more detail.*

Example 6.25

Most modern devices would not work without the products that come from crude oil.

If scientists discovered alcohol today, the government would probably restrict it in the same way as cocaine.

- *Use 'would be able to' instead of 'could' to express a hypothetical ability.*

Example 6.26

Before: If nuclear power were 100% safe, more countries could convince their people to use it.

After: If nuclear power had a perfect safety record, more countries would be able to convince their people to use it.

- *Use 'should'/'should not' instead of 'ought to'/'ought not' to describe an expected or recommended behaviour or situation.*

Example 6.27

Healthcare workers should listen carefully to their patients.

Laboratory personnel should not ignore relevant safety procedures.

- *Use 'must' to express necessity or requirement. This function is easy and straightforward for L2 readers.*

Example 6.28

A specialist nurse must have extensive knowledge about the dialysis machine and how it works to ensure that it operates properly.

You must handle the laboratory rats very carefully to avoid contamination.

- *Use 'must not'* to express prohibition.

Example 6.29

Carcinogens <u>must not be used</u> as solvents or reagents in the last stages
of a drug synthesis.

- *Avoid using 'must' to mean probable.* Instead, reword the sentence using
  'probably' to express a high degree of possibility, a logical deduction or con-
  fident assumption.

Example 6.30

Before: Those clouds <u>must mean</u> it will rain later.

After: Those clouds probably <u>mean</u> it will rain later.

- *Avoid past forms of modals, such as 'may have escaped' or 'should have tested'.*
  Instead try rewording sentences without the modals.

Example 6.31

Before: The toxins <u>may have escaped</u> during laboratory procedures.

After: <u>It is possible that</u> the toxins <u>escaped</u> during laboratory procedures.

OR

After: <u>Perhaps</u> the toxins <u>escaped</u> during laboratory procedures.

Example 6.32

Before: The pharmaceutical company <u>should have tested</u> the product before
distribution.

After: Although it is a requirement, the pharmaceutical company <u>did not test</u> the
product before distribution.

- *Use direct commands for polite requests and permission.* 'Can', 'could', 'may' and 'would' are all used to express polite requests or permission. Usually these are used in spoken communication, but they may also be needed for written instructions. In this case, use direct commands instead. Reserve 'can', 'could', 'may' and 'would' for the other functions as recommended above. These are more common in scientific writing.

Example 6.33

Before: You <u>can</u> spend five minutes on this activity.

After: <u>Spend</u> no more than five minutes on this activity.

Before: You <u>may not use</u> a calculator.

After: You <u>are not allowed to use</u> a calculator. OR <u>Do not use</u> a calculator.

6.6   Limit Clause Per Sentence Ratio

Aim to write sentences that contain only one or two clauses. The fewer clauses per sentence in a text, the higher the readability. Limiting the number of clauses will also help keep your sentences short. Writing guides recommend keeping sentences to fewer than 30 words [2, 3].

Most of the time it is possible to break a multi-clause sentence into individual sentences of one or two clauses. If not, write the sentence as clearly as possible. Example 6.34 shows how to reduce the number of clauses—indicated with square brackets—by making more, but shorter, sentences. A bonus is that the revision results in fewer words overall.

Example 6.34

Before: A simple rule [that can be used to determine the jurisdictional body] [that controls a particular aspect of mining] is to consider [that state laws regulate] [how the mining operation is practiced], [whereas federal government regulates the production, marketing, and distribution of the products]. (1 sentence, 6 clauses, 43 words)

After: There is a simple way to determine [who controls a particular aspect of mining]. State laws control mining operations. Federal laws control the production, marketing, and distribution of the products. (3 sentences of only 1 or 2 clauses each, 29 words)

Another way of clarifying a sentence with multiple clauses is to use bullet points, as shown in Example 6.35.

Example 6.35

Before: Although physics is a fundamental science, it also successfully explains how to build a better steam engine, how to place a satellite in orbit, and how to use energy stored in atomic nuclei to light cities.

After: Although physics is a fundamental science, it also successfully explains how to:

- build a better steam engine;
- place a satellite in orbit;
- use energy stored in atomic nuclei to light cities.

6.7   De-Clutter Clauses

Avoid packing too much into a clause. Grammatically, a single clause can have:

- one or more subjects;
- one or more verbs relating to the same subject(s);
- and one or more objects relating to the same verb(s);
- any number of phrases, such as 'for several years', 'in each of the cells'.

If a single clause has multiple subjects, verbs and objects it can become a complicated jumble. Too many phrases can also make a sentence unwieldy, if not unreadable. Examples 6.36 and 6.37 shows how to create simpler, shorter sentences using the same information.

To de-clutter such clauses, try rewording and breaking up the information into shorter sentences, as in the following two examples.

Example 6.36

Before: During the past few decades, national space agencies from around the world and university planetary sciences departments have collected and processed a large amount of detailed facts, photographs and other information about the functioning of Earth, our complex dynamic planet.

After: During the past few decades, international space agencies and university planetary sciences departments have gathered information about Earth. They have collected and processed a large amount of detailed facts and photographs about our complex dynamic planet.

Example 6.37

Before: The Body Mass Index is not a perfect measurement and does not account for differences in frame size, gender, or muscle mass, and in fact, mis-classifies as many as 25% of people by distinguishing between lean muscle mass and body fat.

After: The Body Mass Index is not a perfect measurement. It does not account for differences in frame size, gender, or muscle mass. In fact, it mis-classifies as many as 25% of people because it does not distinguish between lean muscle mass and body fat.

6.8   Choose Easy Clause Connectors

Choose the most common and the least ambiguous clause connectors. Table 6.2 and Table 6.3 summarise which connectors to use and which to avoid.

Table 6.2 Connectors for compound sentences

Use	Avoid
• and • but • or/nor • so (meaning 'as a result')	• for (meaning 'because') • yet

Table 6.3 Connectors for complex sentences

Use	Avoid
• because • so that • although or though (choose one and use consistently) • which, that • who/whose • where (referring to place) • if/unless • when/as soon as • before, after, until • while (indicating two actions happening at the same time) • since (relating to time)	• as (showing reason) • as (indicating two actions happening at the same time) • since (showing a reason) • where (indicating a certain case or position) • whereas • while (showing contrast

Suggestions for using clause connectors:

- Use 'because' instead of 'for' in a complex sentence. Reserve 'for' for its more common use as a preposition (e.g. medicine for diabetes).

Example 6.38

Before: The discovery was interesting <u>for</u> it offered a general solution to the problem.

After: The discovery was interesting <u>because</u> it offered a general solution to the problem.

- Use 'but' or 'although' instead of 'yet'. Reserve 'yet' for its more common use as an adverb (e.g. not completed yet).

Example 6.39

Before: There is not much that science can do about Nature, <u>yet</u> it has certainly tried.

After: There is not much that science can do about Nature, <u>but</u> it has certainly tried.

- Use 'because' instead of 'as' (showing reason). Reserve 'as' for its use for comparisons (e.g. as big as) or to introduce a short example (e.g. such as).

Example 6.40

Before: <u>As</u> the product is sensitive to light, a dark room is needed to isolate it.

After: <u>Because</u> the product is sensitive to light, a dark room is needed to isolate it.

- Use 'while' instead of 'as' (indicating two actions happening at the same time).

Example 6.41

Before: <u>As</u> you add the acid, stir the mixture constantly.

After: <u>While</u> you add the acid, stir the mixture constantly.

- Use 'because' instead of 'since' (showing a reason). Reserve 'since' for its more common and more easily understood use relating to time.

Example 6.42

Before: Since pesticides and herbicides can affect biological systems, we can classify them as drugs.

After: Because pesticides and herbicides can affect biological systems, we can classify them as drugs.

- Use 'if' or 'when' instead of 'where' (indicating a certain case or position). Reserve 'where' to refer to a place.

Example 6.43

Before: Where there is more than one rational explanation for the experimental observations, some confusion may arise.

After: If there is more than one rational explanation for the experimental observations, some confusion may arise.

- Express contrast with either 'but' in a compound sentence or 'although' in a complex sentence instead of 'whereas'.

Example 6.44

Before: Regular tissue has a limited number of cell divisions, whereas tumour tissue can divide without limit.

After: Regular tissue has a limited number of cell divisions, but tumour tissue can divide without limit.

- Use 'although' instead of 'while' (showing contrast). Reserve 'while' to indicate two actions happening at the same time.

Example 6.45

Before: While these chemicals are toxic, they are safe in the hands of an expert.

After: Although these chemicals are toxic, they are safe in the hands of an expert.

6.9   Leave 'That' in!

Native English speakers commonly omit words such as 'that', 'which' and 'who' in certain types of sentences. Although such sentences are perfectly grammatical and seem more concise, they are more difficult for L2 readers to read.

Clause connectors help L2 readers to distinguish the clauses in a sentence. They also help reduce ambiguity. Examples 6.46–6.51 show the different types of sentences that benefit from leaving in 'that', 'which' or 'who'.

Example 6.46

Before: The gauge shows the tank is empty.

After: The gauge shows that the tank is empty.

Example 6.47

Before: Regular tissue grown in a culture has a limited number of cell divisions.

After: Regular tissue that is grown in a culture has a limited number of cell divisions.

Example 6.48

Before: A bowling ball rolling at constant velocity is in equilibrium until it experiences a non-zero net force.

After: A bowling ball that is rolling at constant velocity is in equilibrium until it experiences a non-zero net force.

Example 6.49

Before: The book of scientific principles <u>published in 1902</u> remained popular with historians.

After: The book of scientific principles <u>that</u> <u>was published in 1902</u> remained popular with historians.

Example 6.50

Before: The people <u>depending on fishing</u> have seen much progress in fish nursery protection.

After: The people <u>who</u> <u>depend on fishing</u> have seen much progress in fish nursery protection.

Example 6.51

Before: Foresters do not know how to create guidelines <u>which</u> <u>villagers in developing countries can use to manage p antations</u>.

After: Foresters do not know how to create gu delines <u>which villagers in developing countries can use to manage plantations</u>.

In each of the above examples, both sentences are correct; both have the same meaning. But in each case, the longer sentence is clearer.

6.10   Paraphrase Hypothetical 'If' Clauses

Present and future conditional sentences with 'if' are generally easy for L2 readers to understand.

Example 6.52

If you boil any liquid, it evaporates.

If you measure the value of k at various temperatures T, then you can calculate other unknown parameters.

However, if the conditional sentence expresses something unreal or contrary to fact, such as the one in Example 6.53, it is much more difficult.

Example 6.53

If fish <u>had</u> larger brains, they <u>would learn</u> how to read.

The first half of the sentence has a past tense verb, but its meaning is not in the past. The meaning of the word 'had' is that the opposite is true. In other words, the fish do *not* have larger brains. The second half of the sentence indicates an imaginary result of the first clause: fish learning to read. This is a confusing concept for many L2 readers.

If fish had larger brains, they would learn to read.

Example 6.54 shows a couple of other ways of expressing this information.

Example 6.54

Imagine a fish with a large brain. This fish would be able to learn to read.

OR

Fish do not have large brains, so they cannot learn to read.

There is no one way of rewording such hypothetical sentences to clarify the meaning. Each hypothetical sentence needs to be thought out and carefully explained. Example 6.55 shows how a completely different wording is needed to avoid a hypothetical sentence.

Example 6.55

Before: If the drug company <u>were to source</u> its raw material only from the bark of the Yew tree, then the Yew tree <u>would become</u> extinct.

After: To prevent the extinction of the Yew tree species, the drug company must find another source of raw material besides the Yew tree bark.

References

1. K. Sainani, C. Elliott, D. Harwell, *Active vs. Passive Voice in Scientific Writing.* [Online presentation] (2015), https://www.acs.org/content/acs/en/acs-webinars/professional-development/active-passive.html
2. A.E. Greene, *Writing Science in Plain English.* Chicago Guides to Writing, Editing, and Publishing (University of Chicago Press, Chicago, 2013)
3. ASD, *International Specification for the Preparation of Technical Documentation in a Controlled Language* (AeroSpace and Defence Industries Association of Europe, Brussels, Belgium, 2017), p. 382

Chapter 7
Go with the Flow

Contents

Abstract Writers use a variety of techniques to organise sentences into a cohesive text that flows logically. These techniques vary from language to language and culture to culture. The cultural differences in techniques affect the ability of some L2 readers to follow a text efficiently. This chapter provides the following guidelines for making texts more transparently cohesive.

Keywords Readable writing · Readable STEM · Writing clarity · Writing STEM

7.1 Give the Main Idea in the First Sentence of a Paragraph

L2 readers typically learn in English reading classes that the first, or sometimes second, sentence in a paragraph provides the main idea. They also learn to read the first sentence of every paragraph as a way of skimming a text, a common technique for preparing them to read more closely.

L2 readers are also often taught to write the main idea in the first sentence. As a result, they come to expect that each paragraph in an English text will have new or different information in it.

In Example 7.1, both paragraphs are the same, except that the second one begins with a topic sentence that provides readers with an orientation to the overall topic.

Example 7.1

Before: A stroke is caused by plaque deposits inside the walls of the major artery to the brain. The plaque reduces blood flow. An ultrasonic generator on the neck can determine the rate of blood flow by reflecting from the red blood cells as they move through the artery. The change in the frequency of ultrasound can detect the speed of red blood cells as they flow through the artery. This provides information about the speed of blood flow and helps doctors diagnose problems such as blood clots.

After: The Doppler effect can be used to assess the risk of a stroke. A stroke is caused by plaque deposits inside the walls of the major artery to the brain. The plaque reduces blood flow. An ultrasonic generator on the neck can determine the rate of blood flow by reflecting from the red blood cells as they move through the artery. The change in the frequency of ultrasound can detect the speed of red blood cells as they flow through the artery. This provides information about the speed of blood flow and helps doctors diagnose problems such as blood clots.

7.2 Give Information Gradually

Once you have the main idea in the first sentence, give information in the following sentences gradually. As much as possible, make each sentence contain only one thought or piece of information. Too much information too quickly makes the text too dense. Dense text forces readers to move through the text slowly. Sometimes they have to reread parts of the text to decipher the meaning.

In Example 7.2, the first paragraph begins with a lot of information packed in the first sentence. The second one, the revision, teases out each piece of information and provides information gradually so that readers can follow it more easily.

Example 7.2

Before: The consequences of not applying human factors, or of wrongly applying human factors through inappropriate methodology, can increase risks of ill health and injury, dissatisfaction and discomfort for workers. For a company, the consequences at the least can be a loss of competitiveness, in terms of productivity, quality, flexibility and timeliness.

After: Risks to workers can increase if human factors are not applied or are wrongly applied. These risks include ill health and injury, dissatisfaction and discomfort for the workforce. These problems can cause companies to lose competitiveness. Companies may lose productivity, quality, flexibility and timeliness.

In Example 7.3, the two texts provide the same information and even begin with the same sentence. But the second text provides the information more gradually.

Example 7.3

Before: Environmental science is the systematic study of our environment and our place in it. Because of the complexity of environmental problems, environmental science depends on many fields of knowledge such as biology, chemistry, earth science, and geography in order to gather all the important information. In addition, the social sciences and humanities, running from political science and economics through art and literature, contribute to an understanding as to how society at large reacts to environmental crises and opportunities. The practice of environmental science is generally targeted to try to understand problems so that solutions in public health and environmental quality can be offered. (103 words, 4 sentences)

After: Environmental science is the systematic study of our environment and our place in it. Environmental issues are complex. Therefore, environmental science depends on many fields of knowledge. Sciences such as biology, chemistry, earth science and geography all contribute important information to environmental science. Social sciences and humanities also contribute important information. For example, political science, economics, art and literature help us understand how society reacts to environmental crises and opportunities. Environmental scientists aim to understand problems and provide solutions to those problems. They aim to solve problems in public health and environmental quality. (93 words, 8 sentences)

7.3   Increase Word Overlap

Repeat key words and phrases as much as possible across sentences and paragraphs in a text. Avoid writing the same ideas using different words or phrases. Repeating key words helps show the relationship between sentences and between different parts of a text. It also helps L2 readers connect ideas in a text, thereby increasing their reading comprehension and speed [1].

Example 7.4 shows how we can increase the word overlap by repeating the word 'minerals'.

Example 7.4

Before: Minerals are all solid at the earth's surface temperature. This matter has orderly crystal structures.

After: Minerals are all solid at the earth's surface temperature. These minerals have orderly crystal structures.

In Example 7.5, we can increase the word overlap by repeating the key words 'Sun' and 'energy' in the second, third and fourth sentences. This makes it easier for L2 readers to connect the parts to the whole mentioned in the first sentence.

Example 7.5

Before: Radiant energy from the Sun reaches the Earth as electromagnetic waves. Part of it is absorbed. Another part is reflected. The remainder goes past the Earth.

After: Radiant energy from the Sun reaches the Earth as electromagnetic waves. Part of the Sun's energy is absorbed. Another part of the energy is reflected. The remainder of the Sun's energy goes past the Earth.

In the longer Example 7.6, we can increase the word overlap by repeating several of these key words and phrases throughout: 'process', 'drug', 'design and development', 'many people', 'skills' and 'technologies'.

Example 7.6

Before: The process of drug design and development involves a large number of complex issues. Therefore, no one person could possibly manage all of the tasks required to discover, develop, and successfuly bring a new therapeutic entity to market. A multidimensional approach is needed to coordinate the efforts of individuals with a wide array of expertise such as medicinal chemistry, in vitro biology, drug metabolism, animal pharmacology, formulations science, process chemistry, clinical research, intellectual property, and many other fields. Enabling technologies, such as high-throughput screening, molecular modelling, pharmaceutical profiling and biomarker studies, also play key roles in modern drug research.

After: New drug design and development is a complex process. Therefore, the process requires many people to discover, develop, and successfully bring a new drug to market. This process of drug design and development must be multidimensional. It requires the combined skills of many people with different expertise. These skills include medicinal chemistry, in vitro biology, drug metabolism, animal pharmacology, formulations science, process chemistry, clinical research, intellectual property and many other fields. Technologies that assist the overall process include high-throughput screening, molecular modelling, pharmaceutical profiling and biomarker studies. These technologies are also important in modern drug research.

7.4   Clarify Substitute Word Reference

Substitute words are words such as 'it', 'they', 'one', 'this', 'here' and 'so'. They refer to other words in the text. Make it crystal clear what these substitute words refer to. Clarifying substitute word reference:

- eliminates ambiguity;
- removes the need for the reader to make inferences;
- reduces the pressure on L2 readers' limited vocabulary.

When the pigs left their pens, Farmer Brown painted them.

If a substitute word can possibly refer to more than one noun or phrase, replace it with the correct noun. Repetition is better than forcing readers to waste time trying to make the connections.

In Example 7.7, the pronouns 'they' and 'them' could possibly refer to 'theoretical models' or 'studies'. We must infer which one. We know it refers to 'theoretical models' because 'studies' cannot be proven or unproven.

However, for many readers, it takes additional time and concentration to make this inference. A simple way to clarify it is to repeat the word 'models'. This eliminates the need to infer, allowing readers to move through the text faster and more easily.

Example 7.7

Before: The theoretical models emerge from various experimental studies of the system. Mostly unproven, they tend to reflect the major schools of thought at the time. A historical timeline shows them in neat progressive development, but the facts emerged in a more random fashion.

After: The theoretical models emerge from various experimental studies of the system. Mostly unproven, these models tend to reflect the major schools of thought at the time. A historical timeline shows these models in neat progressive development, but the facts emerged in a more random fashion.

In Example 7.8, L2 readers must connect the substitute word 'This', in the second sentence, to one of three noun phrases in the first sentence: 'the most recent accomplishment', 'the treatment of infection' or 'the human immunodeficiency virus'. This is especially difficult to do if the words in these phrases are unknown to the reader. We can clarify what 'This' refers to by simply repeating the word 'virus' in the second sentence.

Example 7.8

Before: The most recent accomplishment is the treatment of infection with the human immunodeficiency virus (HIV). This leads to acquired immune deficiency syndrome (AIDS).

After: The most recent accomplishment is the treatment of infection with the human immunodeficiency virus (HIV). This virus leads to acquired immune deficiency syndrome (AIDS).

Avoid using substitute words such as 'so' to refer to an entire phrase or sentence. Instead, use a paraphrase that repeats the key words, as shown in Example 7.9.

Example 7.9

Before: The English scientist, John Dalton, first proposed the atomic theory to account for the experimental observations reported at the time. By doing so, he was able to carry out the earliest predictive studies of new compounds.

After: The English scientist, John Dalton, first proposed the atomic theory to account for the experimental observations reported at the time. By proposing the atomic theory, he was able to carry out the earliest predictive studies of new compounds.

If possible, avoid using 'this', 'that', 'these' and 'those' to refer to entire clauses or sentences. In Example 7.10, 'this problem' refers to an entire situation described in the previous two sentences. We can clarify the text by rewriting the last sentence without the phrase 'this problem'.

Example 7.10

Before: Scientists do not yet have the ability to control the differentiation of an organism's cells. They cannot manipulate cells or insert them into another organism to provide tissue regrowth. The solution to <u>this problem</u> would provide a cure for many medical problems.

After: Scientists do not yet have the ability to control the differentiation of an organism's cells. They cannot manipulate cells or insert them into another organism to provide tissue regrowth. If scientists find a way to control and manipulate cells, they may also find a cure for many medical problems.

7.5   Keep Sentence Types Consistent

Use a limited set of sentence types and use them repeatedly. Use similar wording and similar sentence types for the same types of information so that readers can rely on finding certain types of information in certain parts of a text.

Uniform and consistent sentence constructions are easier for L2 readers to process [1]. While in the course of reading a sentence, readers form expectations of how it will end. Thus, the more predictable the grammatical structures are in a text, the lower the cognitive demand on the reader. This, in turn, gives the reader a chance to concentrate more on meaning. Reiterated sentence structures are cognitively less demanding [1].

Example 7.11 shows a text with several inconsistencies.

Example 7.11

Before:

Para 1 Electrical safety is important for both people and property. For example, in household circuits, as more and more electrical appliances are turned on, the current increases, and the wires become increasingly hotter. This is dangerous. It might start a fire in the house.

Para 2 There are two types of safety devices for electrical circuits: fuses and circuit breakers. Fuses are used mainly in older homes, but newer homes have circuit breakers instead of fuses to protect against fire.

Para 3 The fuse, which is shown in the circuit diagram in Fig. 1, is a safety device that prevents the wires from getting too hot and possibly starting a fire. When the amount of current becomes too high, the fuse filament gets so hot that it melts and breaks the circuit.

Para 4 Two kinds of fuses are usually used in household circuits. The Edison-base fuse has a base with threads similar to the threads on a light bulb (Fig. 2). Thus, they fit into any socket. This can create a problem. Someone might screw a 30-A fuse, for example, into a socket that should have a 15-A fuse. This can be dangerous, so the Type S fuses are often used instead (see Fig. 3). Type S fuses have a threaded adapter that is specific to each size of fuse. Fuses that have a different rating have different threads, and therefore a 30-A fuse cannot be screwed into a 15-A socket. Household fuses are being phased out. However, some automobile and other circuits still have small fuses.

Para 5 In new homes, circuit breakers are used instead of fuses. Circuit breakers serve the same function as fuses. A thermal type of circuit breaker, as illustrated in Fig. 4 (on page 103), uses a bimetallic strip. As the current increases through the strip, it becomes warmer and bends. The strip bends more and then breaks the circuit when it becomes too hot.

Some of the inconsistencies in this text are:

- The phrase 'For example' appears at the beginning of the sentence in paragraph 1, but in the middle of the sentence in paragraph 4. The better style is the one in paragraph 1 with the phrase at the beginning of the sentence where it is clearly visible.
- The second sentence in paragraph 1 is awkward and meandering compared to the other sentences in the paragraph.
- In paragraph 2, the second sentence has two clauses that are unnecessarily of different styles. The first clause begins with the type of device and then matches it to the type of home; the second clause begins with the type of home and then matches it to the device.
- In paragraph 4, sentences 3, 6 and 8 show result in three different ways. Sentence 3 begins with 'Thus' to express a consequence of the previous sentence or clause. Sentence 6 uses 'so' to join the result to its cause. Sentence 8 shows result with 'therefore' embedded in the middle of the sentence. The best style is sentence 3 because the result is clearly signalled at the beginning of the sentence.
- Paragraphs 3, 4 and 5 contain four different ways of directing the reader's attention to diagrams. Use only one way consistently throughout a text.
- Paragraphs 2 and 4 both introduce two elements, but do so in different styles ('There are two types of ...' and 'Two kinds of fuses are ...'). Either style is fine, but they should be similar.
- The last two sentences of paragraph 5 are unnecessarily different in style. Both are complex sentences using connectors ('as' and 'when' respectively). However, the second last sentence begins with the connector and the last sentence puts the connector in the middle of the sentence between its two clauses.

Example 7.12 shows the same text without these inconsistencies.

Example 7.12

After:

Para 1 Electrical safety is important for both people and property. For example, as you turn on more electrical appliances in a house, the current increases. As the current increases, the wires become increasingly hot. This is dangerous because it might start a fire in the house.

Para 2 There are two types of safety devices for electrical circuits: fuses and circuit breakers. Both types of devices protect the home from fire. Older homes usually have fuses, but newer homes have circuit breakers instead.

Para 3 The fuse is a safety device that prevents the wires from getting too hot and possibly starting a fire. See a diagram of the circuit in Fig. 1. When the amount of current becomes too high, the fuse filament gets so hot that it melts and breaks the circuit.

Para 4 There are usually two kinds of fuses in household circuits: the Edison-base fuse and the Type S fuse. See a diagram of the Edison-base fuse in Fig. 2. This fuse has a base with threads similar to the threads on a light bulb. Thus, they fit into any socket. This can create a problem. For example, someone might screw a 30-A fuse into a socket that should have a 15-A fuse. This can be dangerous, so the Type S fuses are often used instead. See a diagram of Type S fuse in Fig. 3. Type S fuses have a threaded adapter that is specific to each size of fuse. Fuses that have a different rating have different threads. Thus, a 30-A fuse cannot be screwed into a 15-A socket. Household fuses are being phased out. However, some automobile and other circuits still have small fuses.

Para 5 In new homes, circuit breakers are used instead of fuses. They serve the same function as fuses. A thermal type of circuit breaker uses a bimetallic strip. See a diagram of a thermal circuit breaker in Fig. 4. As the current increases through the strip, it becomes warmer and bends. When it becomes too hot, the strip bends more and breaks the circuit.

7.6   Use Signposts

Signposts are words or short phrases that show readers the relationship between sentences and between different parts of a text.

Expressions such as 'First', 'However', and 'As a result' help L2 readers move more easily through a text. L2 readers usually learn the most common signposts in English writing classes.

Put the signpost at the beginning of the sentence where L2 readers can see it clearly, as shown in Example 7.13. Follow it with a comma to further distinguish it. Use a limited number of signposts repeatedly and consistently.

Example 7.13

Before: The atmosphere has always trapped heat near the earth's surface. Human activities such as the burning of fossil fuels have, <u>however</u>, contributed greatly to the increase in greenhouse gas levels.

After: The atmosphere has always trapped heat near the earth's surface. <u>However</u>, human activities such as the burning of fossil fuels have contributed greatly to the increase in greenhouse gas levels.

Signposts serve many different functions, such as showing contrast, introducing an example, or indicating a result. For each function, choose the most common expression and use it consistently.

Table 7.1 shows suggested signposts for a variety of functions. All of the recommended signpost words are among the 3,000 most common words in English.

Table 7.1 Signpost suggestions for various functions

To indicate	Use	Instead of
A full-sentence example	For example	For instance A case in point Namely To demonstrate As an illustration
Chronological order	First Second numbered/bulleted points Initially Then Next Eventually Finally	Firstly Secondly Third, Fourth, et cetera At first After that Subsequently Last(ly)
Co-occurring events	At the same time or Simultaneously	Meanwhile
An additional idea	In addition or Also (sometimes needs to be mid-sentence)	Further Furthermore Moreover
An opposite idea or contrast	However or But[a] or Instead	On the other hand Otherwise Nonetheless Nevertheless In contrast By comparison Still (reserve for its other uses) Conversely (except for use in mathematics)
Similarity	Similarly	Likewise
Emphasis	In fact	Indeed On the contrary (negative)
An effect or result	As a result or Therefore or Thus	Accordingly As a consequence Consequently Hence Then Because of this

[a]Although it was once forbidden to use 'But', 'Or' or 'And' to start a sentence, it has become increasingly accepted and is perfectly clear to the reader.

7.7   Minimise Referrals

Avoid referring to earlier or later parts of the text with expressions such as 'as seen earlier', 'as explained below' and 'as discussed in a later chapter.' Reading is primarily a linear process, and the referral practice can distract readers. Instead, consider repeating the material rather than sending the reader to other parts of the text [2].

Referring to later parts of a text is acceptable if it is simply an assurance that something will be explained more fully at a later point.

References

1. S.A. Crossley, J. Greenfield, D.S. McNamara, Assessing text readability using cognitively based indices. Tesol Q. **42**(3), 475–493 (2008)
2. E.H. Weiss, Reducing burdens, in *The Elements of International English Style* (M. E. Sharpe, New York, London, 2005), p. 84

Chapter 8
Balance Your Style

Contents

Abstract Style, or register, refers to the degree of formality of writing. Styles of writing range from the very informal, such as a text message to a friend, to the very formal, such as a legal document. Traditionally, science writers have used a highly formal style. Writers, editors and even readers *expected* text to be filled with abstract nouns, passive verbs, ornate sentences and unnecessarily dry descriptions. But this is changing—and for the better. Scientific writers and publishers now recognise that such a style is not appropriate for most of their readers (A.E. Greene, Writing Science in Plain English. Chicago Guides to Writing, Editing, and Publishing (University of Chicago Press, Chicago, 2013), K. Sainani, C. Elliott, D. Harwell, Active vs. Passive Voice in Scientific Writing. [Online presentation] (2015), https://www.acs.org/content/acs/en/acs-webinars/professional-development/active-passive.html [1, 2]). The most appropriate writing style depends on context, purpose and audience. Certainly, an overly formal style is inappropriate for L2 readers. But so too is an overly *in*formal style. Such a style tends to be full of casual language that L2 readers do not normally encounter in an English classroom setting. This chapter provides guidelines for finding the appropriate balance, a 'Goldilocks' style that is neither too formal nor too informal.

Keywords Simple writing · Writing STEM · Clear writing · Writing clarity

8.1 Involve Your Readers

Draw your readers into your text by asking questions and using 'you' and 'we' where appropriate. This encourages readers to use their background knowledge and experience to help them make sense of a text. When they actively participate in the reading experience, they process and understand the text better.

In Example 8.1, the use of questions and the pronoun 'you' invites the reader to compare their own experience with the new information in the text.

Example 8.1

Static friction occurs when there is enough frictional force to stop relative motion between two surfaces. For example, have you ever tried to move a refrigerator by sliding it across the floor? When you push on the side of it, it does not move. Why? Because you have not pushed hard enough. Thus, the force of static friction between the bottom of the refrigerator and the floor is equal or greater than the force you applied. As a result, there is no motion. It is a static condition.

8.2 Balance Concrete and Abstract

Use concrete and specific language as much as the subject matter allows. Supplement abstract and general ideas with concrete and specific examples if possible.

Concrete words refer to things that we can see, hear, smell, taste or touch. Abstract words describe ideas or concepts, things that do not have a physical form. Although abstract and general language is often necessary to describe ideas concisely and in depth, it can make texts unnecessarily vague.

Concrete words are not necessarily easier than abstract words for L2 readers. L2 readers probably have well developed conceptual abilities in their first language and that can help them understand the same concepts in their second language [3]. However, some concepts in their first language may not have corresponding equivalencies in their L2 [4]. In any case, abstract words do tend to make any text more vague.

In Example 8.2, we can increase concrete language by converting three of the abstract words—'identification', 'generation', and 'engineering'—into concrete words. The revision also has fewer words.

Example 8.2

Before: The identification of problems and the generation of solutions are the first steps of electrical engineering. (16 words)

After: Electrical engineers must identify problems first and then find solutions. (10 words)

By converting the word 'engineering' into 'engineers', we are now talking about real people. And those people actively do things, i.e. 'identify' and 'find'.

Example 8.3 shows how we can increase concrete language by using active rather than passive verbs and by converting three abstract nouns into concrete verbs. The revision is a clearer and shorter sentence.

Example 8.3

Before: The heat and work terms <u>were separately analysed</u> by researchers, and methods for the <u>calculation</u>, <u>measurement</u>, and <u>interpretation</u> of each <u>were presented</u>. (22 words)

After: Researchers <u>analysed</u> the heat and work terms. They <u>presented</u> methods for <u>calculating</u>, <u>measuring</u>, and <u>interpreting</u> each. (16 words)

Example 8.4 shows how we can increase the concrete language by getting rid of the vague words and phrases and completely rewriting the text into shorter sentences. The result is one word longer, but is much clearer.

Example 8.4

Before: <u>The characterisation of certain</u> health effects of ionising radiation as having no <u>threshold dose</u> means that all internal and external sources that contribute to the received dose by <u>individual members of the population</u> must <u>be considered</u>. (36 words, one sentence)

After: For some health effects, we do not know if there is a safe maximum dose of ionising radiation. To assess your health risk you need to sum both internal and external amounts of your exposure to radiation. (37 words, 2 sentences)

Example 8.5 shows how we can make the text more concrete by making the information more specific and by converting vague verbs to action verbs and abstract nouns to concrete nouns. The revision contains fewer, but more specific words and shorter, but more active sentences.

Example 8.5

Before: In order to consider all aspects of health and safety in the production and distribution of chemicals, one needs to ascertain the size of the problem. Although there is no absolute measure of the scale of global production and use of chemical substances, it is estimated that the average annual world production of such substances is in excess of 450 million tons. Other estimates indicate that there are currently identified over five million distinct chemical compounds, with this number increasing at the rate of over a third of a million per year. Many of these compounds are clearly not in everyday commercial or industrial use, but it is estimated that at least 100,000 chemical substances can be considered to be in everyday use on a substantial scale, and that this number is being added to at the rate of at least several hundred per year, in the case of substances which are produced in quantities in excess of one ton per year. (162 words, 4 sentences)

After: Rules and regulations help prevent accidents in workplaces that produce and use chemicals. To make these rules, we need to know the size of the problem. How many chemical substances are produced? How much of those substances are produced? We do not know the exact number of chemical substances that are produced and used in the world. However, estimates indicate that more than five million different chemical compounds and more than 450 million tons of chemicals are produced. The number of chemical compounds increases by more than 300,000 every year. However, most of these compounds are not commonly used on a large scale. Only about 100,000 chemical substances are produced and used in quantities of more than one ton. The number of these substances increases by several hundred every year. (130 words, 10 sentences)

8.3   Balance Friendly and Folksy

Use a simple, but not a conversational style. A folksy style, such as that used by the For Dummies [5] and The Complete Idiot's Guide [6] series, is suitable for L1 readers who are struggling intellectually with the material. But for L2 readers, whose challenge is the language and not necessarily the concepts, a folksy style presents more problems than it solves.

A folksy style is based on conversational language. It tends to be wordy and inefficient. In addition to more words for L2 readers to trawl through, there are more idiomatic or metaphoric expressions and more culturally foreign references that are normally used only in casual conversation. L2 readers who have learned English in mainly formal settings will not have had exposure to this kind of casual language.

Example 8.6 contains idiomatic and overly casual expressions that make the paragraph difficult for L2 readers. We can maintain a friendly tone without using those expressions, and make the paragraph much shorter in the process.

Example 8.6

Before: Physicists always have a way of getting their heads into the craziest places. Most of the time, those places involve really big or really small numbers. Suppose you're dealing with the distance between the Sun and Pluto. That's about 5,890,000,000 metres. You have a whole bunch of metres on your hands! And they come with a whole bunch of zeroes. But don't worry. Physics has a way of dealing with really large and really small numbers. We use something called scientific notation to help make our numbers look tidier and more digestible. (92 words)

After: In physics, we use very large and very small numbers. For example, the distance between the Sun and Pluto is about 5,890,000,000 metres. That may seem like a lot of zeroes. But physics has a way of dealing with very large and very small numbers. We use scientific notation to help make numbers easier to read and understand quickly. (59 words)

In Example 8.7, we see a cultural reference that may not resonate with readers who are not American. In the revision of this paragraph, we can retain the analogy, which is an excellent way to explain complex ideas, without making the cultural reference. We can also replace the idiomatic expressions with more universally understood phrases.

Example 8.7

Before: We can think of atoms as people. Most people are ordinary working stiffs but some are independently rich fat cats. Most ordinary people want to be independently rich. In the same way, ordinary atoms want to be noble gases, such as helium [He], neon [Ne], xenon [Xe] and argon [Ar]. These elements are found in the last column of the Periodic table. The other atoms try to imitate the noble gases, which are the Rockefellers and Vanderbilts of atoms.

After: We can think of atoms as people. Most people work at ordinary jobs to make money, but some people do not have to work because they come from rich families. Most ordinary people want to be independently rich. In the same way, ordinary atoms want to be noble gases, such as helium [He], neon [Ne], xenon [Xe] and argon [Ar]. These elements are found in the last column of the Periodic table. The other atoms try to imitate the noble gases, which are the stable and independent atoms.

Before using analogies and specific examples, first consider the extent to which they are relevant and universally understood.

8.4   Avoid Writing About Writing

Avoid adding comments about what you are writing, such as:

- It would be best if we start with …
- Perhaps the most important information is …
- As will be seen in the next section …
- As we discussed in the last chapter …
- It is important to know …
- We cannot talk about [something] without considering …

Such expressions are usually unnecessary. They merely add extra verbiage for readers to trawl through.

References

1. A.E. Greene, *Writing Science in Plain English*. Chicago Guides to Writing, Editing, and Publishing (University of Chicago Press, Chicago, 2013)
2. K. Sainani, C. Elliott, D. Harwell, *Active vs. Passive Voice in Scientific Writing*. [Online presentation]   (2015),   https://www.acs.org/content/acs/en/acs-webinars/professional-development/active-passive.html
3. B. Laufer, G.C. Ravenshorst-Kalovski, Lexical threshold revisited: lexical text coverage, learners' vocabulary size and reading comprehension. Read. Foreign Lang. **22**(1), 15–30 (2010)
4. N.C. Ellis, Vocabulary acquisition: word structure, collocation, word-class, and meaning, in *Vocabulary: Description, Acquisition and Pedagogy*, ed. by N. Schmitt, M. McCarthy (Cambridge University Press, Cambridge, 1997), pp. 122–139
5. Dummies, https://www.dummies.com/
6. Complete Idiots Guides, http://www.penguin.com/static/html/aboutus/adult/alpha

Chapter 9
Clear the Page

Contents

Abstract Layout, how content is distributed on the page, has a huge impact on L2 readability. Writers and publishers can greatly enhance readability by paying attention to some key layout features. Unfortunately, modern printing practices often end up reducing overall readability in an attempt to save paper (Weiss in The Elements of International English Style. M. E. Sharpe, New York; London, 2004 [1]). Layout issues are especially problematic for L2 readers whose first language is typographically different from English. For example, languages that read from right to left, such as Arabic, and non-alphabet-based languages, such as Chinese and Japanese, look very different on the page. This makes clarity of page layout critical for many L2 readers. This chapter discusses how to reduce distractions on the page to enhance readability for L2 readers.

Keywords Clear writing layout · Readable layout · Improved text layout

9.1 Use Left-Aligned Single Columns

Use a single-column format and align the text left. Do not justify it.

A multiple-column format is distracting, especially if it is justified. Justified text spreads out the words on each line to give straight edges on both sides of the column. It may look neat and tidy to L1 readers, but it can cause confusion for L2 readers because it produces awkward gaps in a line of text. Example 9.1 shows the effect of irregular gaps on the lines of justified text.

Example 9.1

Before:
Physics is a primary fundamental science that studies the basic principles of the universe. By the 1990s, it seemed that nearly all of Nature could be accounted for by experimentation and intellectual application. A century of the particle view of light had been changed to a wave view in the early 1800's. Maxwell's electromagnetic theory allowed a range of radiation frequencies outside of the visible to be predicted. The conservation laws of energy, momentum, angular momentum, and charge were well established. The fundamental forces of gravity were well researched and	understood. Atomic theory of matter, finally accepted in the early 1900's, proposed atoms as the smallest unit of matter. These atoms were related to specific elements. Molecules are built from atoms, and these can come from different elements. The kinetic theory of gases, in turn, relies on atomic theory. The period near the turn of the 19th century saw four key discoveries that set the foundations for the atomic age. X-rays, radioactivity, the electron, and the splitting of spectral lines were all reported. These and many other research avenues form the basis of modern physics.

Hyphenation solves the gap problem, but it causes another problem. Example 9.2 shows how justification with hyphenation breaks up words in ways that make them more difficult for L2 readers to recognise.

Example 9.2

> *Before:*
>
> Physics is a primary fundamental science that studies the basic principles of the universe. By the 1990s, it seemed that nearly all of Nature could be accounted for by experimentation and intellectual application.
>
> A century of the particle view of light had been changed to a wave view in the early 1800's. Maxwell's electromagnetic theory allowed a range of radiation frequencies outside of the visible to be predicted.
>
> The conservation laws of energy, momentum, angular momentum, and charge were well established. The fundamental forces of gravity were well researched and understood.
>
> Atomic theory of matter, finally accepted in the early 1900's, proposed atoms as the smallest unit of matter. These atoms were related to specific elements. Molecules are built from atoms, and these can come from different elements. The kinetic theory of gases, in turn, relies on atomic theory.
>
> The period near the turn of the 19th century saw four key discoveries that set the foundations for the atomic age. X-rays, radioactivity, the electron, and the splitting of spectral lines were all reported. These and many other research avenues form the basis of modern physics.

A multiple-column format can be even more confusing when visual inserts are awkwardly inserted.

A left-aligned single-column format with no hyphenation provides readers with the most readable flow and is easier to scan quickly for specific information.

Example 9.3

After:

Physics is a primary fundamental science that studies the basic principles of the universe. By the 1990's, it seemed that nearly all of Nature could be accounted for by experimentation and intellectual application.

A century of the particle view of light had been changed to a wave view in the early 1800's. Maxwell's electromagnetic theory allowed a range of radiation frequencies outside of the visible to be predicted.

The conservation laws of energy, momentum, angular momentum, and charge were well established. The fundamental forces of gravity were well researched and understood.

Atomic theory of matter, finally accepted in the early 1900's, proposed atoms as the smallest unit of matter. These atoms were related to specific elements. Molecules are built from atoms, and these can come from different elements. The kinetic theory of gases, in turn, relies on atomic theory.

The period near the turn of the 19th century saw four key discoveries that set the foundations for the atomic age. X-rays, radioactivity, the electron, and the splitting of spectral lines were all reported. These and many other research avenues form the basis of modern physics.

Left alignment is especially important for L2 readers whose first language is read from right to left. For them, reading from left to right feels unnatural, especially when they are grappling with complex material. By left justifying the text, you provide a clue to these readers about the correct direction of the text—i.e. start from the straight edge.

9.2   Choose Clear Fonts and Spacing

Clear fonts and spacing are especially important for L2 readers whose first language is not based on the Latin alphabet.

Use clear, simple fonts of a reasonable size. Standard, sans serif fonts work best. Sans serif fonts are increasingly used for online documents because of their simplicity. Example 9.4 shows the effects of changing a serif font 10 to a sans serif font 12.

Example 9.4

> *Before:*
> The Earth is a solid, spherical, rocky body with oceans and an atmosphere. The Earth is unique in our solar system. It is the only planet with large amounts of water on the surface, an atmosphere with oxygen and a temperate climate.
>
> *After:*
> The Earth is a solid, spherical, rocky body with oceans and an atmosphere. The Earth is unique in our solar system. It is the only planet with large amounts of water on the surface, an atmosphere with oxygen and a temperate climate.

Generous letter and line spacing also help increase readability. In Example 9.5, both texts are the same font type and size, but the revision has increased letter and line spacing.

Example 9.5

> *Before:*
> The Earth is a solid, spherical, rocky body with oceans and an atmosphere. The Earth is unique in our solar system. It is the only planet with large amounts of water on the surface, an atmosphere with oxygen and a temperate climate.
>
> *After:*
> The Earth is a solid, spherical, rocky body with oceans and an atmosphere. The Earth is unique in our solar system. It is the only planet with large amounts of water on the surface, an atmosphere with oxygen and a temperate climate.

9.3   Minimise Typographical Features

Several typographical features are available for indicating emphasis in a text, such as:

- **bold**
- *italics*
- CAPITAL LETTERS
- coloured letters
- larger fonts
- shading
- parentheses, brackets and quotation marks—e.g. ( ). [ ], { }, " ", ' '.

Choose a limited number of these and use them consistently.

Use bold, larger fonts or colour to highlight important information. These features draw readers' attention to key information, even if they are just skimming the text.

Avoid using italics or all capital letters. These two features are harder for L2 readers to process.

Also, limit the use of parentheses to the absolute essentials, such as abbreviations. Information in parentheses may cause breaks in reading continuity. Use other ways to draw attention to certain types of information. For example, to mark definitions and alternative meanings, etc. choose boldface or colour, if available.

If possible, avoid using quotation marks. The misuse of quotation marks may lead to confusion, and they should be reserved strictly for quotations. There are different American and British conventions that are followed when using quotation marks. This can present confusion for both L1 and L2 readers.

9.4   Use Shorter Paragraphs

Write shorter paragraphs with spaces between them. Separating paragraphs with a line, rather than distinguishing them with an initial indentation, creates more white space. Increased white space improves readability by reducing the visual density of the page. This, in turn, reduces strain and encourages readers to keep going.

As shown in Example 9.6, breaking up large pieces of dense text also makes it less likely for readers to overlook important information.

Example 9.6

Before:
Sounds that have intensity levels of more than 120 dB can harm our ears. Even short exposures to high-intensity noise can tear your eardrums. Damaged eardrums cause permanent hearing loss. But sounds that have lower intensity levels can also damage your eardrums if you are exposed to them over a long period of time. For example, motorcycles and loud rock concerts can harm your hearing. Loud sounds over a long period of time can damage the hearing nerves and the sensitive hair cells of your inner ear. Signs of hearing damage include not hearing speech clearly, feeling pressure in the ear, and hearing a ringing sound in your ear. If the damage is not severe, these symptoms may go away in a few minutes or a few hours after exposure to the sounds.

After:
Sounds that have intensity levels of more than 120 dB can harm our ears. Even short exposures to high-intensity noise can tear your eardrums. Damaged eardrums cause permanent hearing loss.

But sounds that have lower intensity levels can also damage your eardrums if you are exposed to them over a long period of time. For example, motorcycles and loud rock concerts can harm your hearing. Loud sounds over a long period of time can damage the hearing nerves and the sensitive hair cells of your inner ear.

Signs of hearing damage include not hearing speech clearly, feeling pressure in the ear, and hearing a ringing sound in your ear. If the damage is not severe, these symptoms may go away in a few minutes or a few hours after exposure to the sounds.

9.5   Use Bulleted Lists

Use bulleted lists to present information that would otherwise be embedded—and buried—in a paragraph. This is a good way to highlight key information. Use a numbered list if the order of items is important.

Using bulleted lists is a quick and easy way to increase readability dramatically. They are clearer for L2 readers than an extended narrative that contains several key points. Bulleting is most effective when each of the bullet points is at the same priority level. Sub-bulleting can increase complexity and decrease clarity.

Example 9.7

Before:

A mechanism is a detailed proposal that shows electron movements to explain all bond breaking and making. The main information shown in a mechanism includes any reactive intermediates or transition states involved, curly arrows showing all electron movements, information about relative reaction rates in stepwise reactions and any relevant stereochemical features.

After:

A mechanism is a detailed proposal that shows electron movements to explain all bond breaking and making. The main information shown in a mechanism includes:

- any reactive intermediates or transition states involved;
- curly arrows showing all electron movements;
- information about relative reaction rates in stepwise reactions;
- any relevant stereochemical features.

9.6   Use Headings and Subheadings

Guide the L2 reader through the text using headings and subheadings. This allows the L2 reader to process information more readily and effectively. Example 9.8 shows how headings can make information more prominent and meaningful.

Example 9.8

Before:

Explaining Evolution

It is easy to observe and record the changes in the evolution process. But it is much harder to understand how evolution occurs. Populations change over time. New species develop through natural selection. Selection occurs in individuals, but we see the results in a population over a period of time.

Populations consist of individuals that share the same genetic composition. Changes in a gene or chromosome can cause one individual to have a new characteristic. This is called a mutation. Random mutations cause very small differences when the genes pass from parents to their children. The survival of any individual depends on how well it is suited to its living environment.

Individuals that have better genetic composition have a better chance of surviving and passing on these good genes to their children. As a result, these good genetic characteristics spread in the population. In this way, the population adapts to its environment. A well-adapted species tends to survive, but a badly adapted species die out.

Evolution is the result of random variations and varying probabilities for reproduction. Thus, evolution is unpredictable. Also, natural selection depends on the raw materials available. It cannot create something out of nothing, and occurs in small steps. For example, insect wings did not appear in a single step in the children of a parent without wings. Instead, insect wings developed little by little, step by step, over several generations.

Each little step of development depends on natural selection. Eventually, one individual insect was born with wings, a random mutation. This insect was more likely to survive because it could find food and escape its enemies more easily.

After:

How Evolution Occurs

It is easy to observe and record the changes in the evolution process. But it is much harder to understand how evolution occurs. Populations change over time. New species develop through natural selection. Selection occurs in individuals, but we see the results in a population over a period of time.

Mutation process

Populations consist of individuals that share the same genetic composition. Changes in a gene or chromosome can cause one individual to have a new characteristic. This is called a mutation. Random mutations cause very small differences when the genes pass from parents to their children. The survival of any individual depends on how well it is suited to its living environment.

Genetics controls survival

Individuals that have better genetic composition have a better chance of surviving and passing on these good genes to their children. As a result, these good genetic characteristics spread in the population. In this way, the population adapts to its environment. A well-adapted species tends to survive, but a badly adapted species die out.

Random variations

Evolution is the result of random variations and varying probabilities for reproduction. Thus evolution is unpredictable. Also, natural selection depends on the raw materials available. It cannot create something out of nothing, and occurs in small steps. For example, insect wings did not appear in a single step in the children of a parent without wings. Instead, insect wings developed little by little, step by step, over several generations.

Each little step of development depends on natural selection. Eventually, one individual insect was born with wings, a random mutation. This insect was more likely to survive because it could find food and escape its enemies more easily.

9.7   Choose Visuals with Impact

Use visuals that help the reader understand the text. Avoid using gratuitous photos, diagrams or tables. Limit non-essential, general visuals that do not specifically enhance the text. Visuals that do not support or enhance the text tend to distract L2 readers.

Label visuals clearly and refer to them in the appropriate place(s) in the text. Position them close to their in-text reference. This helps the L2 readers connect the content to the visual.

If at all possible, avoid wrapping the text tightly around the visual. White space around the visual helps reduce the density of the page.

In Example 9.9, two of the pictures contribute nothing significant to the comprehension of the text. Also, the details of the visual become a jumble because the text is wrapped tightly around the visual.

Example 9.9

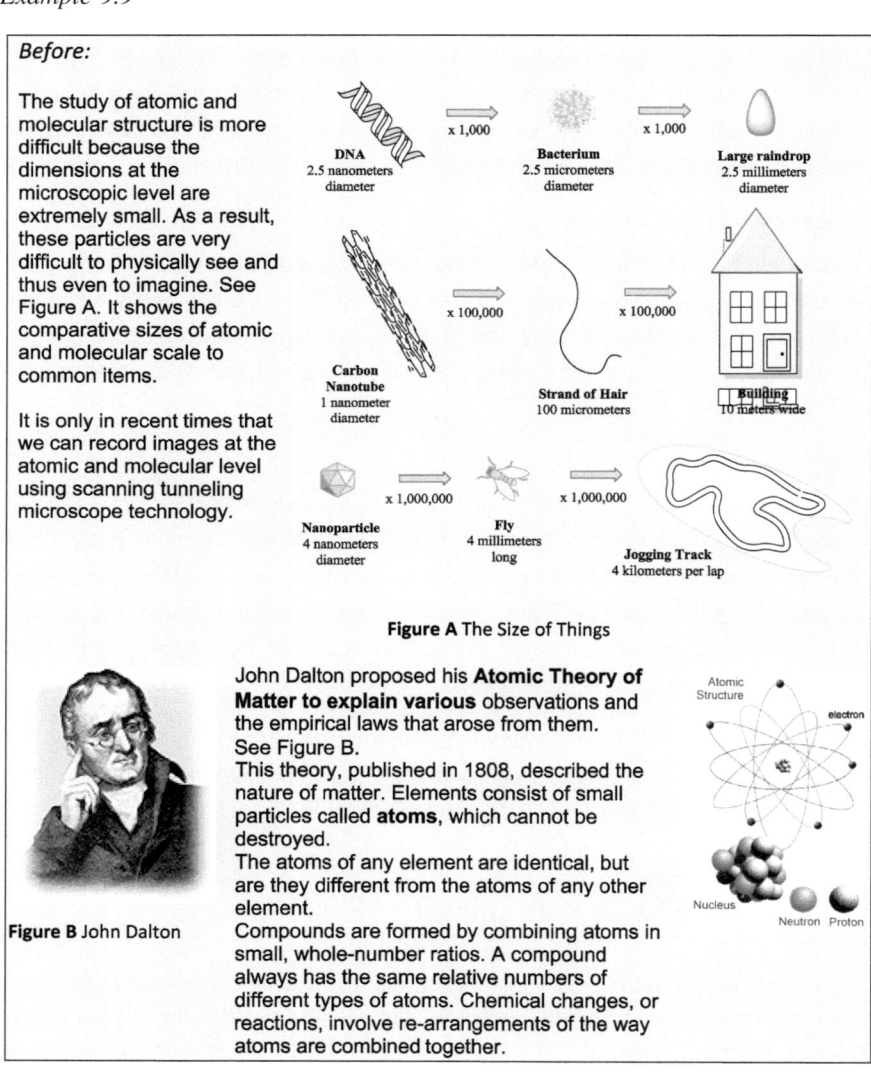

Before:

The study of atomic and molecular structure is more difficult because the dimensions at the microscopic level are extremely small. As a result, these particles are very difficult to physically see and thus even to imagine. See Figure A. It shows the comparative sizes of atomic and molecular scale to common items.

It is only in recent times that we can record images at the atomic and molecular level using scanning tunneling microscope technology.

DNA
2.5 nanometers
diameter

x 1,000

Bacterium
2.5 micrometers
diameter

x 1,000

Large raindrop
2.5 millimeters
diameter

Carbon
Nanotube
1 nanometer
diameter

x 100,000

Strand of Hair
100 micrometers

x 100,000

Building
10 meters wide

Nanoparticle
4 nanometers
diameter

x 1,000,000

Fly
4 millimeters
long

x 1,000,000

Jogging Track
4 kilometers per lap

Figure A The Size of Things

John Dalton proposed his **Atomic Theory of Matter to explain various** observations and the empirical laws that arose from them. See Figure B.
This theory, published in 1808, described the nature of matter. Elements consist of small particles called **atoms**, which cannot be destroyed.
The atoms of any element are identical, but are they different from the atoms of any other element.
Compounds are formed by combining atoms in small, whole-number ratios. A compound always has the same relative numbers of different types of atoms. Chemical changes, or reactions, involve re-arrangements of the way atoms are combined together.

Atomic
Structure

electron

Nucleus

Neutron   Proton

Figure B John Dalton

In the following revision, the two irrelevant pictures are removed and the remaining visual is enlarged and re-positioned to give a cleaner look with more white space. In addition, some of the text has been converted to bulleted points for further clarity.

After:

The study of atomic and molecular structure is more difficult because the dimensions at the microscopic level are extremely small. As a result, these particles are very difficult to physically see and thus even to imagine. See Figure A. It shows the comparative sizes of atomic and molecular scale to common items.

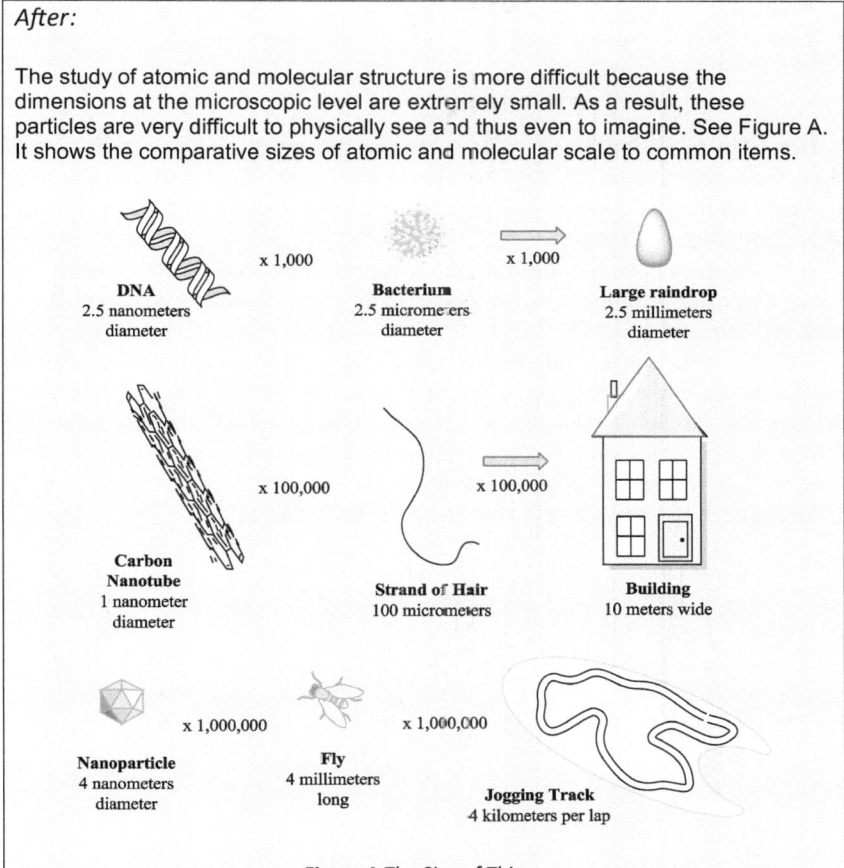

Figure A The Size of Things

It is only in recent times that we can record images at the atomic and molecular level using scanning tunneling microscope technology. John Dalton proposed his **Atomic Theory of Matter to explain various** observations and the empirical laws that arose from them. This theory, published in 1808, described the nature of matter.

- Elements consist of small particles called **atoms**, which are indestructible.
- The atoms of any element are identical, but are they different from the atoms of any other element.
- Compounds are formed by combining atoms in small, whole-number ratios.
- A compound always has the same relative numbers of different types of atoms.
- Chemical changes, or reactions, involve re-arrangements of the way atoms are combined together.

9.8 Minimise Special Features

Be very selective about the inclusion of non-essential special features such as:

- sidebars;
- boxed special topics discussions, such as examples of real world applications;
- end-of-chapter reviews;
- extraneous pictures and artwork.

Consider collecting extra features together at the end of a section or chapter rather than breaking up the page within the main text. Much additional material is available on the internet or other external reference sources and is not essential for inclusion within the text.

In an effort to capture the attention of readers, modern science materials tend to overuse special features, each needing their own introduction to explain their purpose, use, and benefits.

For L2 readers, many of these special features often confuse and disrupt the reading flow. While trying to cope with the main material, they are also forced to discern the essential from the extraneous. Thus, the extra(neous) material can detract from their comprehension of the main material. This is an unnecessary frustration in an already challenging task.

Example 9.10 shows a text with unnecessary sidebars.

Example 9.10

Before:

Science is a way of thinking about the world around us. It is a way to get answers to questions and to solve problems. It is a way of finding out what we do not know by first seeing what we do know. It is more correct to think of science as a method - a **scientific method.**

Ibn Al Haytham (Alhazen), 965—1039, was a polymath in Iraq. He is often called the father of modern scientific method because he emphasized the importance of experimental data and reproducibility of its results.

See Figure 1 showing the process of the scientific method. It is a guide to asking and answering questions by making observations and doing experiments. Use it to search for cause and effect relationships in Nature. Scientific method has a series of steps. But, new information or thinking might cause the scientist to back up and repeat steps at any point in the process. A process like scientific method that involves such repetition is an **iterative process.**

Galileo Galilei, 1564-1642. Albert Einstein said, "All knowledge of reality starts from experience and ends in it. Propositions arrived at by purely logical means are completely empty as regards reality. Because Galileo saw this, and particularly because he drummed it into the scientific world, he is the father of modern physics—indeed, of modern science altogether."

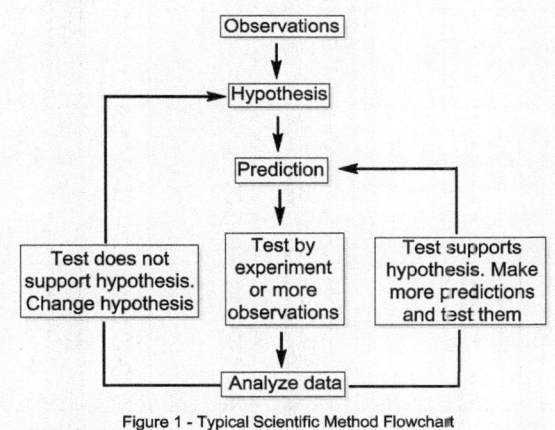

Figure 1 - Typical Scientific Method Flowchart

Although the sidebars are interesting, they distract from the key information in the main text, and may be removed to enhance reading focus.

After:

Science is a way of thinking about the world around us. It is a way to get answers to questions and to solve problems. It is a way of finding out what we do not know by first seeing what we do know. It is more correct to think of science as a method - a **scientific method.**

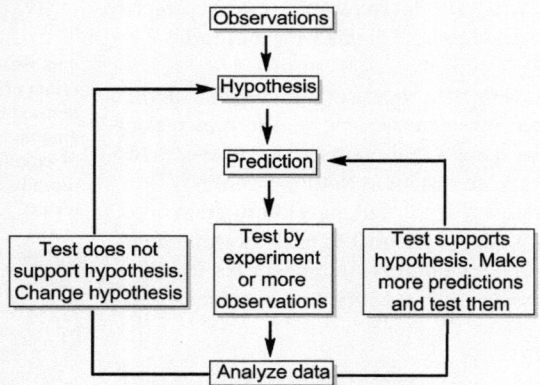

Figure 1 - Typical Scientific Method Flowchart

See Figure 1 showing the process of the scientific method. It is a guide to asking and answering questions by making observations and doing experiments. Use it to search for cause and effect relationships in Nature. Scientific method has a series of steps. But, new information or thinking might cause the scientist to back up and repeat steps at any point in the process. A process like scientific method that involves such repetition is an **iterative process.**

Reference

1. E.H. Weiss, *The Elements of International English Style* (M. E. Sharpe, New York, London, 2005)

Chapter 10
Tools for Readability

Contents

Abstract Advances in the field of computational linguistics have provided some useful tools to help you assess and improve text readability (Vajjala and Meurers in The 7th workshop on the innovative use of NLP for building educational applications. Association for Computational Linguistics, Canada, 2012 [1]). By taking much of the guesswork out of the writing and editing process, these tools can help you make texts more accessible and more comprehensible for L2 readers. They are not prescriptive guides; they merely provide measurements and feedback to help identify areas of text difficulty. This chapter discusses three such tools that are appropriate and convenient for writing and editing texts for L2 readers.

Keywords Measure readability · Simplified English · Improved readability

10.1 Compleat Web VP [2]

Compleat Web VP is an online vocabulary profiler that helps you to easily identify and reduce the level of difficulty of the vocabulary in your text. It analyses the vocabulary in your text and categorises each word in the text into frequency levels. It starts with K-1, the most common 1,000 words in the English language, and runs to K-25. There is also an Off-List category that shows names, specialised technical vocabulary and highly unusual words.

Compleat Web VP can be accessed at https://www.lextutor.ca/vp/comp/. Figure 10.1 shows the main web page of this tool.

© The Author(s), under exclusive license to Springer Nature Singapore Pte Ltd. 2019   109
C. Roos and G. Roos, *Real Science in Clear English*, SpringerBriefs
in Education, https://doi.org/10.1007/978-981-13-7820-1_10

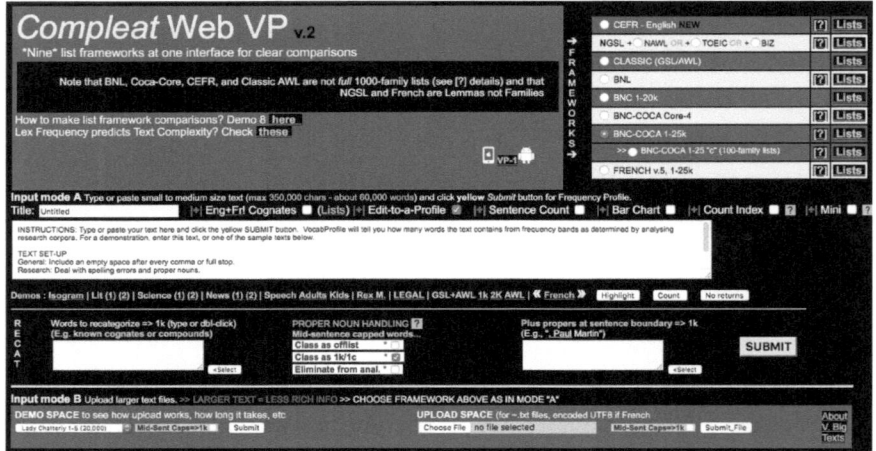

Fig. 10.1 Compleat Web VP web page

Compleat Web VP is robust and easy to use. The following steps show how to use this tool to improve the readability of a text.

10.1.1 Choose a Framework

Select a framework in the top right corner of the web page, as shown in Fig. 10.1. The most suitable framework for a wider academic application is the BNC-COCA-25. Other frameworks are designed for teachers and students learning vocabulary.

The BNC-COCA-25 provides 25 frequency levels of 1000 words each. It is an international frequency list that harmonises the British National Corpus (BNC) and the Corpus of Contemporary English (COCA). Both lists are based on extensive corpora drawn from academic texts as well as spoken texts, fiction, popular magazines, and newspapers. The BNC and COCA complement each other well.

The BNC-COCA 25 also includes the Academic Word List (AWL) in the first 3,000 words. The AWL is a 570-word list of non-technical words that are considered essential for reading academic texts [3].

10.1.2 Enter the Text into the Tool

Remove any visuals in your text and submit it. There are two ways to do this.

- Use 'Input Mode A' to paste your text directly into the window at the centre of the page. Then click on 'SUBMIT'.
- For files in excess of 400,000 characters, use 'Input Mode B' at the bottom of the page to upload your text. It must be in ∼.txt format. Then click on 'Submit File'.

Figure 10.2 shows a short text entered into the profiler using Input Mode A. Entering a title in the box above the text is optional.

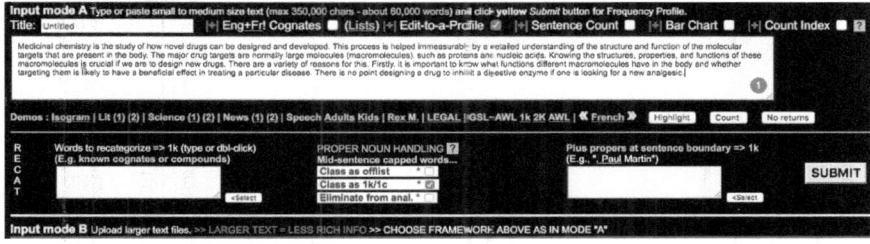

Fig. 10.2 Example of text entered into Compleat Web VP (input mode A)

We will use the short text in Example 10.1 to show how to use this vocabulary profiler to improve the readability for L2 readers.

Example 10.1

Medicinal Chemistry

Medicinal chemistry is the study of how novel drugs can be designed and developed. This process is helped immeasurably by a detailed understanding of the structure and function of the molecular targets that are present in the body. The major drug targets are normally large molecules (macromolecules) such as proteins and nucleic acids. Knowing the structures, properties, and functions of these macromolecules is crucial if we are to design new drugs. There are a variety of reasons for this. First, it is important to know what functions different macromolecules have in the body and whether targeting them is likely to have a beneficial effect in treating a particular disease. There is no point designing a drug to inhibit a digestive enzyme if one is looking for a new analgesic.

10.1.3 Analyse the Results

After submitting the file, you will get an instant breakdown of the percentage of your text that falls within each level. The information is broken down in different ways according to word families, types, and tokens. Figure 10.3 shows a screenshot of the vocabulary profile of the Medicinal Chemistry text.

Freq. Level	Families (%)	Types (%)	Tokens (%)	Cumul. token %
K-1 Words :	44 (62.0)	49 (59.04)	92 (71.3)	71.3
K-2 Words :	10 (14.1)	12 (14.46)	12 (9.3)	80.6
K-3 Words :	8 (11.3)	12 (14.46)	14 (10.9)	91.5
K-4 Words :	5 (7.0)	5 (6.02)	5 (3.9)	95.4
		Coverage 95 ▣		
K-5 Words :	1 (1.4)	1 (1.20)	1 (0.8)	96.2
K-6 Words :				
K-7 Words :				
K-8 Words :				
K-9 Words :				
K-10 Words :				
K-11 Words :	1 (1.4)	1 (1.20)	1 (0.8)	97.0
K-12 Words :	1 (1.4)	1 (1.20)	1 (0.8)	97.8
		Coverage 98		
K-13 Words :				
K-14 Words :				
K-15 Words :				
K-16 Words :				
K-17 Words :	1 (1.4)	1 (1.20)	3 (2.3)	100.0
K-18 Words :				
K-19 Words :				
K-20 Words :				
K-21 Words :				

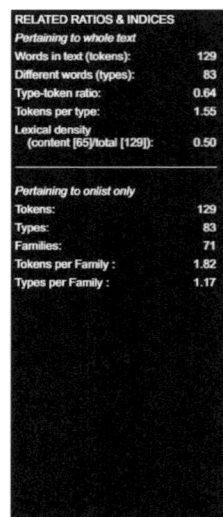

RELATED RATIOS & INDICES	
Pertaining to whole text	
Words in text (tokens):	129
Different words (types):	83
Type-token ratio:	0.64
Tokens per type:	1.55
Lexical density (content [65]/total [129]):	0.50
Pertaining to onlist only	
Tokens:	129
Types:	83
Families:	71
Tokens per Family :	1.82
Types per Family :	1.17

Fig. 10.3 Vocabulary profile of medicinal chemistry text

Figure 10.4 shows a close-up of the left box of Fig. 10.3, as that is the focus of the analysis. It shows five columns of information.

- Frequency Level indicates categories of words according to their frequency in English. K-1 words are the highest frequency; K-25 (not shown) is the lowest frequency. Below K-25 (also not shown) is a category called 'Off-List'. These are words that do not appear on any list, usually names or highly technical words.
  In this example, 95% of the words fall into the 4,000 most frequent words (K-4). 98% coverage is not achieved until K-12, the 12,000 most frequent words in English.
  You will notice that most of the vocabulary of any given text falls within the K-1 level. This is because the short non-content words, such as 'a', 'the', 'in', and 'of' make up a large part of any text. The next K-levels tend to decrease in percentage because the number of shorter, non-content words decreases.
- Families (%) provides the number and percentage of different word families in each category. A word family is a group of related words, such as 'go, goes, went, gone, going' and 'read, readable, unreadable, readability'. The appearance of 'go' and 'went' in a text would be counted as one word.
- Types (%) provides the number and percentage of different words—not families of words—in each category. The appearance of 'go', 'goes' and 'went' in a text would be counted as three words.
- Tokens (%) provides the number and percentage of all words in each category. In our example, a total of 11 words fall outside of the 3,000 most common words—5 in K-4, 1 in each of K-5, K-11 and K-12 and 3 in K-17.

Freq. Level	Families (%)	Types (%)	Tokens (%)	Cumul. token %
K-1 Words :	44 (62.0]	49 (59.04]	92 (71.3)	71.3
K-2 Words :	10 (14.1]	12 (14.46]	12 (9.3)	80.6
K-3 Words :	8 (11.3)	12 (14.46]	14 (10.9)	91.5
K-4 Words :	5 (7.0)	5 (6.02)	5 (3.9)	95.4
Coverage 95 🔲				
K-5 Words :	1 (1.4)	1 (1.20)	1 (0.8)	96.2
K-6 Words :				
K-7 Words :				
K-8 Words :				
K-9 Words :				
K-10 Words :				
K-11 Words :	1 (1.4)	1 (1.20)	1 (0.8)	97.0
K-12 Words :	1 (1.4)	1 (1.20)	1 (0.8)	97.8
Coverage 98				
K-13 Words :				
K-14 Words :				
K-15 Words :				
K-16 Words :				
K-17 Words :	1 (1.4)	1 (1.20)	3 (2.3)	100.0

Fig. 10.4 Vocabulary profile of medicinal chemistry text (left box enlarged)

- Cumul. Token (%) provides the cumulative percentage of words. This example shows that 98% coverage is achieved within the most common 12,000 words in English. Our recommended aim discussed in Chap. 5, is to reach 98% coverage within the 4,000 (K-4) frequency level in order to achieve good readability.

The black box on the right in Fig. 10.3 shows the relative diversity of the vocabulary in your text. In general, the more words that are repeated in a text, the lower the vocabulary diversity. The lower the vocabulary diversity, the easier it is to read. Therefore, a lower type–token ratio indicates a text that is easier to read.

10.1.4   Use Types List to Reduce the Vocabulary Difficulty

If you scroll down further through the output, you can see the Types List, which clearly shows which words fall into which categories. Figure 10.5 shows the Types List of the Medicinal Chemistry text from K-1 to K-5. There are three words that fall in lower frequency categories not shown in this screenshot: 'analgesic' in K-11, 'nucleic' in K-12, and 'macromolecules' in K-17.

Types List [1]
type_[number of tokens]

BNC-COCA-1,000 types: [fams 44 : types 49 : tokens 92]

[Extract]

a_[7] and_[5] are_[4] as_[1] be_[1] body_[2] by_[1] can_[1] different_[1] drug_[2] drugs_[2] firstly_[1] for_[2] have_[2] helped_[1] how_[1] if_[2] important_[1] in_[3] is_[7] it_[1] know_[1] knowing_[1] large_[1] looking_[1] major_[1] new_[2] no_[1] normally_[1] of_[5] one_[1] particular_[1] point_[1] present_[1] reasons_[1] study_[1] such_[1] that_[1] the_[7] them_[1] there_[2] these_[1] this_[2] to_[4] treating_[1] understanding_[1] we_[1] what_[1] whether_[1]

BNC-COCA-2,000 types: [fams 10 : types 12 : tokens 12]

[Extract]

design_[1] designed_[1] designing_[1] detailed_[1] developed_[1] disease_[1] effect_[1] immeasurably_[1] likely_[1] medicinal_[1] process_[1] properties_[1]

BNC-COCA-3,000 types: [fams 8 : types 12 : tokens 14]

[Extract]

crucial_[1] function_[1] functions_[2] molecular_[1] molecules_[1] novel_[1] proteins_[1] structure_[1] structures_[1] targeting_[1] targets_[2] variety_[1]

BNC-COCA-4,000 types: [fams 5 : types 5 : tokens 5]

[Extract]

acids_[1] beneficial_[1] chemistry_[1] digestive_[1] inhibit_[1]

BNC-COCA-5,000 types: [fams 1 : types 1 : tokens 1]

[Extract]

enzyme_[1]

Fig. 10.5 Types List showing which words of the text fall into each category

We can increase the readability of the Medicinal Chemistry text by ensuring that 98% of the text falls within the K-4 frequency level (as suggested in Chap. 5). There are three ways to achieve this:

- Identify essential technical vocabulary and ensure they are adequately defined in or before the text. Then remove them from the calculations. For example, in the Medicinal Chemistry text, we would consider the terms 'macromolecules' 'enzyme' and 'nucleic acids' as essential terminology. We define 'macromolecules' within the text itself and we would have defined 'nucleic acids' prior to the text. We would not count these three words in the profile percentage.
- Replace higher frequency words with lower frequency words. If there is a word in K-2 that works just as well as one in K-4, choose the former. This will help offset the inclusion of necessary technical words, which add to the reader's cognitive load even if they are defined. An online thesaurus is useful for this process. The quickest way to determine if an alternative word is a higher frequency is to type it into a separate Compleat Web VP window and submit it for analysis. For example, in the Medicinal Chemistry text, we can change:

- 'novel' (K3) to 'new' (K1);
- 'immeasurably' (K-2,) to 'greatly' (K-1); 'immeasurably' appears as a K-2 word because of the root 'measure,' but it is less recognisable with its prefix and suffix. In addition, its use in this text is somewhat non-scientific term and inaccurate;
- 'crucial' (K-3) to 'very important' (K-1);
- 'variety' (K-3) to 'many different' (K-1);
- 'to have a beneficial (K-4) effect in treating' to 'to help (K-1) in the treatment';
- 'inhibit' (K-4) to 'block' (K-2).

• Remove low-frequency non-essential technical words or replace them with higher frequency words. The Types List allows you to examine words in the lower frequency categories and decide if you can replace them with higher frequency alternatives. In the Medicinal Chemistry text, we might decide to change 'analgesic' (K-11) to 'pain-killer' (K-1). 'Painkiller' is usually written as one word, and as such is an Off-List word. But by hyphenating it, as recommended in Sect. 5.9, the components—'pain' and 'killer'—become K1 words, making it easier for L2 readers to recognise the term.

We now have the revised text with changes underlined and essential technical words double-underlined, as shown in Example 10.2.

Example 10.2

Medicinal Chemistry (revised)

Medicinal chemistry is the study of how new drugs can be designed and developed. This process is helped greatly by a detailed understanding of the structure and function of the molecular targets that are present in the body. The major drug targets are normally large molecules (macromolecules) such as proteins and nucleic acids. Knowing the structures, properties, and functions of these macromolecules is very important if we are to design new drugs. There are many different reasons for this. First, it is important to know what functions different macromolecules have in the body and whether targeting them is likely to help in the treatment of a particular disease. There is no point designing a drug to block a digestive enzyme if one is looking for a new pain-killer.

We can now run the revised text through the Compleat Web VP. We can either delete the essential technical words before running it or eliminate them from calculations after.

Figure 10.6 shows the revised Medicinal Chemistry text output. 98% of the text now falls within K-3. The percentage of K1 words has jumped from just over 71% in the original example to over 80%.

Freq. Level	Families (%)	Types (%)	Tokens (%)	Cumul. token %
K-1 Words :	49 (75.4)	55 (70.51)	99 (<u>80.5</u>)	80.5
K-2 Words :	9 (13.8)	11 (14.10)	11 (<u>8.9</u>)	89.4
K-3 Words :	5 (7.7)	9 (11.54)	11 (<u>8.9</u>)	98.3
		Coverage 95 ⁇		
K-4 Words :	2 (3.1)	2 (2.56)	2 (<u>1.6</u>)	99.9
		Coverage 98		

Fig. 10.6   Revised vocabulary profile of the medicinal chemistry text

10.2   Hemingway Editor [4]

Although not specifically designed with L2 readers in mind, Hemingway Editor is a text editor that can help address many of the guidelines for keeping sentence structures simple, as described in Chap. 6. Its strength—as well as its limitations—lies in its simplicity. It is available online or as a reasonably priced desktop application.

Hemingway Editor can be accessed at http://www.hemingwayapp.com. Figure 10.7 shows the home page of this tool.

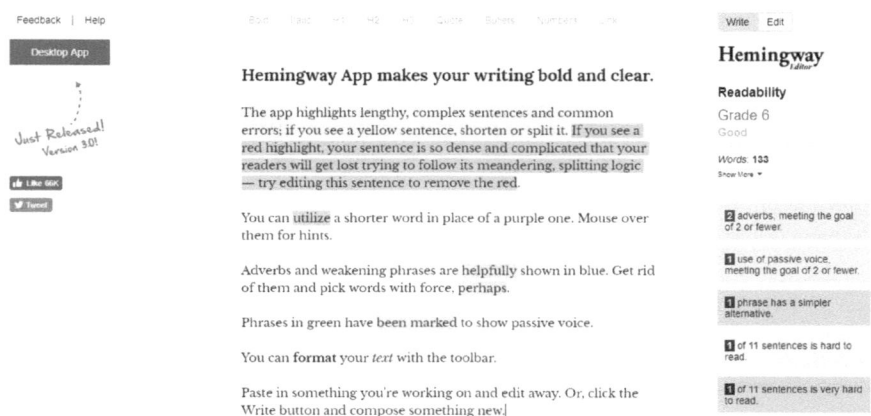

Fig. 10.7   Hemingway Editor home page

Hemingway Editor has six key features. Figure 10.8 shows a close-up view of these features that appear on the right side of the home page.

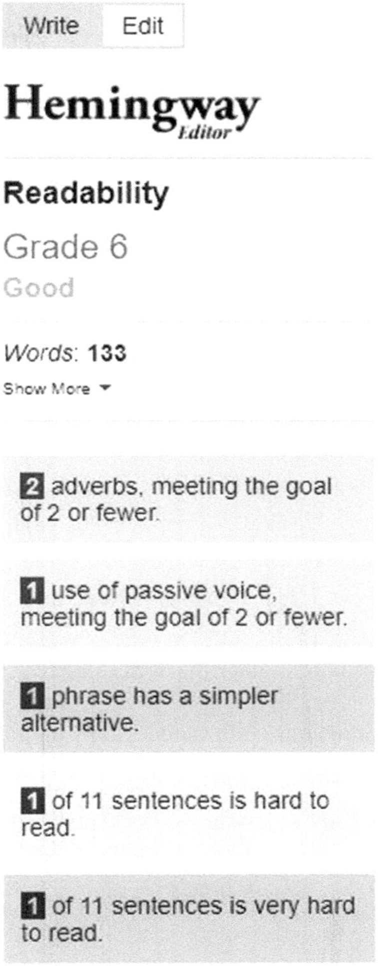

Fig. 10.8 Hemingway
Editor's key features

Write   Edit

Hemingway *Editor*

Readability

Grade 6

Good

Words: **133**

Show More ▼

2 adverbs, meeting the goal of 2 or fewer.

1 use of passive voice, meeting the goal of 2 or fewer.

1 phrase has a simpler alternative.

1 of 11 sentences is hard to read.

1 of 11 sentences is very hard to read.

For our purposes, some features are more useful than others.

- The readability score at the top, expressed as a grade level in the American school system, can provide gratifying results as you edit. However, it is not all that useful for determining text difficulty for L2 readers. It is an outdated and overly simplistic tool originally designed for L1 readers in the US military. Instead, let Coh-Metrix (Sect. 10.3) do the job of determining the readability level. It provides a much more sophisticated and appropriate measure.
- Blue highlights indicate adverbs and 'weakening phrases'. This feature is useful for fiction writers. For example, the Help section of the website advises writers to replace 'walked slowly' with 'tip-toed' or 'crept'. For L2 readers 'walked

slowly' is much easier to understand, even if less poetic. That said, some adverbs are simply not necessary and should be deleted. This feature provides a goal limit based on the length of the text.

- Green highlights indicate passive verbs. This feature also provides a goal limit based on the length of the text, but you need to decide whether the passive verbs in your text are necessary or not.
- Purple highlights indicate words that seem too long. Occasionally, this feature is useful. If you hover over the highlighted word, the application suggests an alternative. For example, 'use' can replace 'utilise'. However, most often the alternative is not sufficiently precise or appropriate. Also, shorter words are not necessarily easier for L2 readers. Use this feature cautiously and rely on Compleat Web VP to properly analyse the vocabulary in your text.
- Yellow and red highlights indicate sentences that are too long. These are the most useful features, along with the green highlights for passive verbs. Yellow and red highlights can help identify overly long noun phrases, cluttered clauses and sentences with too many clauses. But bear in mind that length alone does not make a sentence difficult for L2 readers.

The following steps show how to use Hemingway Editor. You will see how some examples from Chap. 6 illustrate the programme's strengths and limitations for improving readability.

10.2.1   Type or Paste the Text into the Application

Choose either the Write mode or the Edit mode at the top right-hand corner of the page. Write mode fades out the editing tools, allowing you to write without distractions. Edit mode shows the highlighting of problem areas as described above.

10.2.2   Use Red Highlights to Show Long Noun Phrases, Complicated Clauses and Sentences with Too Many Clauses

Hemingway Editor merely highlights sentences that are very long. It is up to you to identify the specific problem and decide how to fix it. For example, Fig. 10.9 shows a sentence in the text editor that is highlighted in red.

Before:
The extensive research and development required to provide the appropriate safety and effectiveness of new drug products is complicated, costly, and time consuming.

Fig. 10.9   Overly long sentence indicated by red highlights

In this case, the overly long sentence is due to its long noun phrases. We will fix that, but we will ignore the green highlights, which erroneously indicate a passive verb. The phrase 'is complicated' is actually an active verb plus an adjective. Figure 10.10 shows a revision of the example with shorter sentences that are now an acceptable length.

After:
We need to do extensive research and development to provide safe and effective drugs. This research and development is a complicated, costly, and time-consuming process.

Fig. 10.10 Revision of overly long sentence into shorter sentences

10.2.3 Use Yellow Highlights Cautiously

Avoid oversimplifying the sentence to satisfy Hemingway Editor. For example, removing words that signal logical connections makes text *more*, not less, difficult to read. Words such as 'but,' 'because,' and 'for example' are important for helping readers understand how the ideas in a text are connected.

Figure 10.11 shows a sentence with red highlights that has been edited into two shorter sentences, which in turn are highlighted yellow.

Before:
The Body Mass Index is not a perfect measurement and does not account for [for differences in frame size, gender, or muscle mass], and in fact mis-classifies as many as 25% of people [by distinguishing between lean muscle mass and body fat].

After:
The Body Mass Index is not a perfect measurement. It does not account for differences in frame size, gender, or muscle mass. In fact, it mis-classifies as many as 25% of people because it does not distinguish between lean muscle mass and body fat.

Fig. 10.11 Overly long sentence converted to two shorter sentences

We can further edit this example to remove the yellow highlights. Figure 10.12 shows this second edit to split the last sentence into two even shorter sentences.

Bold   Italic   H1   H2   H3   Quote   Bullets   Numbers   Link

Before:
The Body Mass Index is not a perfect measurement and does not account for [for differences in frame size, gender, or muscle mass], and in fact mis-classifies as many as 25% of people [by distinguishing between lean muscle mass and body fat].

After:
The Body Mass Index is not a perfect measurement. It does not account for differences in frame size, gender, or muscle mass. In fact, it mis-classifies as many as 25% of people because it does not distinguish between lean muscle mass and body fat.

|After (2nd revision):
The Body Mass Index is not a perfect measurement. It does not account for differences in frame size, gender, or muscle mass. In fact, it mis-classifies as many as 25% of people. It does not distinguish between lean muscle mass and body fat.

Fig. 10.12 Long sentences converted to shorter sentences (second revision)

As you can see, this additional edit makes Hemingway Editor happy—no more yellow highlights. However, we have removed the logical connection between these sentences, the word 'because'. Without this word, the text is more difficult to make sense of.

10.2.4 Use Green Highlights to Check on Passive Verbs

Decide whether the passive verbs are necessary, using the guideline in Sect. 6.4. If not, rewrite the sentences using active verbs. Figure 10.13 shows three sentences with four instances of passive verbs.

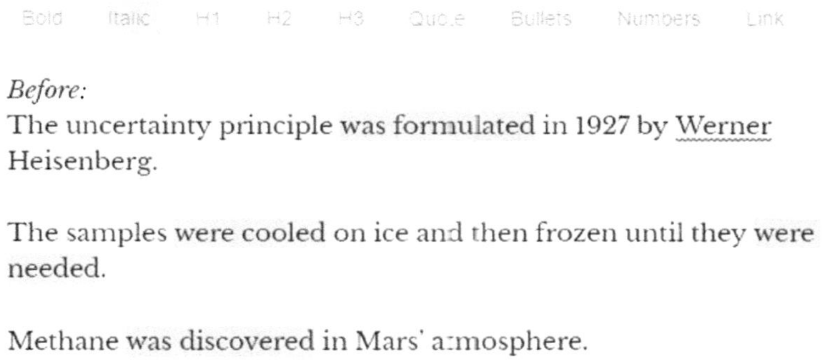

Bold   Italic   H1   H2   H3   Quote   Bullets   Numbers   Link

Before:
The uncertainty principle was formulated in 1927 by Werner
Heisenberg.

The samples were cooled on ice and then frozen until they were
needed.

Methane was discovered in Mars' atmosphere.

Fig. 10.13 Sentences with passive verbs

We can change the first three instances of passive verbs as shown in Fig. 10.14.

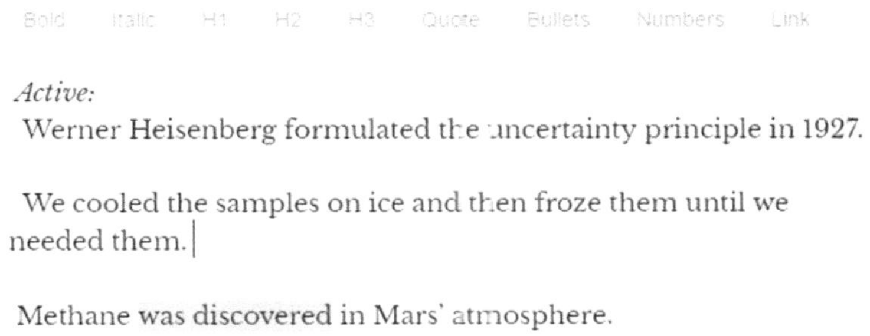

Bold   Italic   H1   H2   H3   Quote   Bullets   Numbers   Link

Active:
Werner Heisenberg formulated the uncertainty principle in 1927.

We cooled the samples on ice and then froze them until we
needed them.

Methane was discovered in Mars' atmosphere.

Fig. 10.14 Revision of sentences with some passive verbs changed to active

We choose not to change the fourth passive verb (third sentence) because we do
not want to shift the focus of the sentence. If we write 'NASA scientists discovered
methane in Mars' atmosphere', we shift the focus from methane to NASA scientists. We want to keep the focus on the discovery of methane.

10.2.5 *Export the Text Using the File Menu*

In the desktop application, you have the option to export (and import) texts using
the File menu.

10.2.6 Use the Help Section for More Information

The Help section describes its features more thoroughly. As the Hemingway Editor developers themselves acknowledge, the editing features are merely suggestions. Nevertheless, its simplicity and instant feedback help break years-old habits with a minimum of fuss.

10.3 Coh-Metrix Index [5]

Coh-Metrix is a computational tool that allows you to gauge the overall reading difficulty of a text. It can be accessed at http://tool.cohmetrix.com/. It works best with the Firefox or Chrome browsers. Figure 10.15 shows the main page of this tool. For background information about the tool, you can go to their home page at http://cohmetrix.com/.

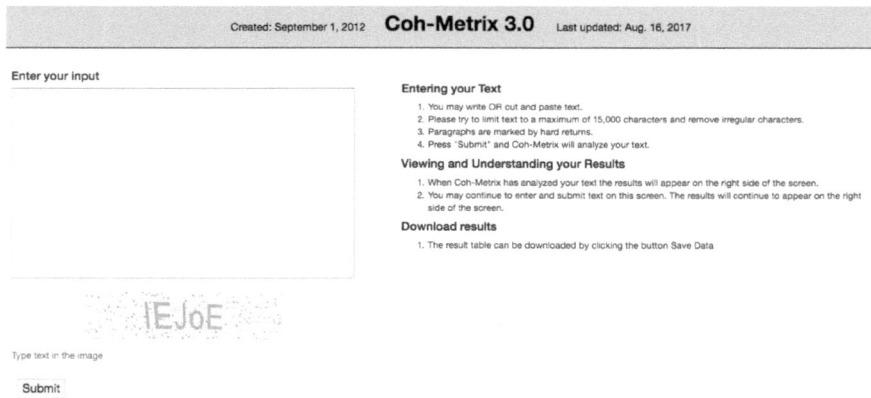

Fig. 10.15 Coh-Metrix computational tool web page

The Coh-Metrix tool is one of the most sophisticated automated text assessment tools available. Other tools, such as the popular Flesch–Kincaid Grade Level, are too narrowly focused on surface-level linguistic features, such as the number of syllables, words, and sentences. Because of their narrow range of features, such tools do not reliably indicate the level of difficulty for L2 readers [6]. Nor do they provide much information to help with editing.

Coh-Metrix, on the other hand, uses a wide range of language, discourse, and conceptual features to measure text difficulty. Its readability formula goes beyond measuring aspects of the words on the page. It reflects how readers actually process a text by taking into account the psycholinguistic and cognitive models of reading. Research on the Coh-Metrix tool indicates that it can accurately assess the difficulty of academic texts [7].

The following steps show how to use Coh-Metrix.

10.3.1 Paste Text into the Screen

You can submit a text of up to 15,000 characters at a time. Then, enter the CAPTCHA code shown below the text and click on submit. The results may take a few minutes to display.

We will use the text in Example 10.3.

Example 10.3

The Process of Drug Design and Development

The process of drug design and development involves a large number of complex issues. Therefore, no one person could possibly manage all of the tasks required to discover, develop, and successfully bring a new therapeutic entity to market. A multidimensional approach is needed to coordinate the efforts of individuals with a wide array of expertise such as medicinal chemistry, in vitro biology, drug metabolism, animal pharmacology, formulations science, process chemistry, clinical research, intellectual property and many other fields. Enabling technologies such as high-throughput screening, molecular modelling, pharmaceutical profiling and biomarker studies, also play key roles in modern drug research.

10.3.2 Analyse the Results

The results will appear on the right side of the screen. Figure 10.16 shows some of the results, indices 1—20 and 104—106. Coh-Metrix has 106 indices, each of which is a measurable indicator of a particular aspect of text difficulty. Index 106 (RDL2) is the most useful one for our purposes. It refers specifically to L2 readability and is based on a formula that combines the measurements:

- *Word overlap.* Measures how often content words overlap between sentences. Increased word overlap improves readability by helping L2 readers link different parts of the text.
- *Word frequency.* L2 readers are more likely to understand high-frequency words than low-frequency ones.
- *Syntactic similarity.* This measures the uniformity and consistency of parallel grammatical constructions in a text. The more uniform the constructions are, the easier the text is for L2 readers to process.

The higher the number on the L2 readability index (RDL2), the more readable the text. Aim to make your text higher than 15. As Fig. 10.16 shows, our example is considerably lower, at only 0.805.

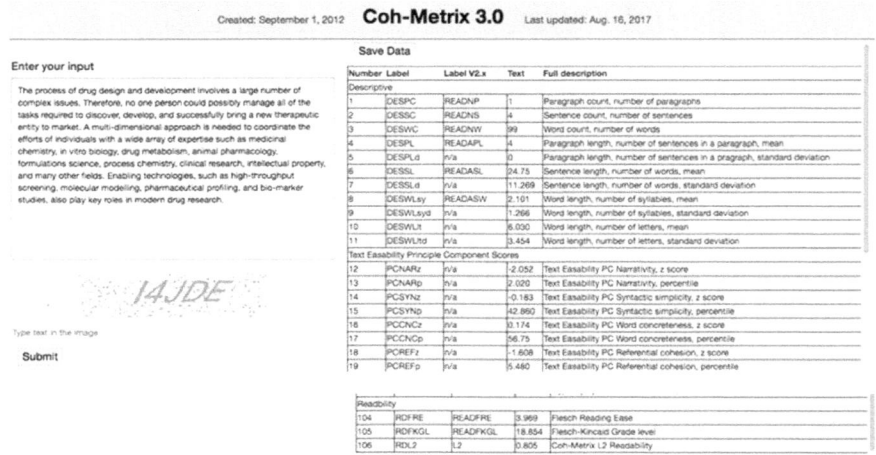

Created: September 1, 2012 **Coh-Metrix 3.0** Last updated: Aug. 16, 2017

Number Label		Label V2.x	Text	Full description
Descriptive				
1	DESPC	READNP	1	Paragraph count, number of paragraphs
2	DESSC	READNS	4	Sentence count, number of sentences
3	DESWC	READNW	99	Word count, number of words
4	DESPL	READAPL	4	Paragraph length, number of sentences in a paragraph, mean
5	DESPLd	n/a	0	Paragraph length, number of sentences in a paragraph, standard deviation
6	DESSL	READASL	24.75	Sentence length, number of words, mean
7	DESSLd	n/a	11.269	Sentence length, number of words, standard deviation
8	DESWLsy	READASW	2.101	Word length, number of syllables, mean
9	DESWLsyd	n/a	1.266	Word length, number of syllables, standard deviation
10	DESWLlt	n/a	6.030	Word length, number of letters, mean
11	DESWLltd	n/a	3.454	Word length, number of letters, standard deviation
Text Easability Principle Component Scores				
12	PCNARz	n/a	-2.052	Text Easability PC Narrativity, z score
13	PCNARp	n/a	2.020	Text Easability PC Narrativity, percentile
14	PCSYNz	n/a	-0.183	Text Easability PC Syntactic simplicity, z score
15	PCSYNp	n/a	42.860	Text Easability PC Syntactic simplicity, percentile
16	PCCNCz	n/a	0.174	Text Easability PC Word concreteness, z score
17	PCCNCp	n/a	56.75	Text Easability PC Word concreteness, percentile
18	PCREFz	n/a	-1.608	Text Easability PC Referential cohesion, z score
19	PCREFp	n/a	5.480	Text Easability PC Referential cohesion, percentile

Readability				
104	RDFRE	READFRE	3.969	Flesch Reading Ease
105	RDFKGL	READFKGL	18.854	Flesch-Kincaid Grade level
106	RDL2	L2	0.805	Coh-Metrix L2 Readability

Enter your input

The process of drug design and development involves a large number of complex issues. Therefore, no one person could possibly manage all of the tasks required to discover, develop, and successfully bring a new therapeutic entity to market. A multi-dimensional approach is needed to coordinate the efforts of individuals with a wide array of expertise such as medicinal chemistry, in vitro biology, drug metabolism, animal pharmacology, formulations science, process chemistry, clinical research, intellectual property, and many other fields. Enabling technologies, such as high-throughput screening, molecular modelling, pharmaceutical profiling, and bio-marker studies, also play key roles in modern drug research.

Type text in the image

Submit

Fig. 10.16 Coh-Metrix analysis results of text

10.3.3   Revise the Text and Resubmit

We will submit the revision as shown in Example 10.4.

Example 10.4

The Process of Drug Design and Development (revised)

New drug design and development is a complex process. Therefore, the process requires many people to discover, develop, and successfully bring a new drug to market. This process of drug design and development must be multidimensional. It requires the combined skills of many people with different expertise. These skills include medicinal chemistry, in vitro biology, drug metabolism, animal pharmacology, formulations science, process chemistry, clinical research, intellectual property and many other fields. Technologies that assist the overall process include high-throughput screening, molecular modelling, pharmaceutical profiling and biomarker studies. These technologies are also important in modern drug research.

10.3.4   Compare the Revised Results with the Original

When you input subsequent texts in the same session, the results appear in new columns next to the original. The columns are labelled Text, Text2, Text3 and so on. This makes it easy to compare results of texts before and after revision. Figure 10.17 shows the results of the revised example text in the Text2 column.

Created: September 1 2012   **Coh-Metrix 3.0**   Last updated: Aug. 16, 2017

Save Data

Number Label		Label ˜2.x	Text	Text2	Full description
Descriptive					
1	DESPC	READMP	1	1	Paragraph count, number of paragraphs
2	DESSC	READMS	4	7	Sentence count, number of sentences
3	DESWC	READMW	99	96	Word count, number of words
4	DESPL	READMPL	4	7	Paragraph length, number of sentences in a paragraph, mean
5	DESPLd	n/a	0	0	Paragraph length, number of sentences in a pragraph, standard deviation
6	DESSL	READMSL	24.75	13.714	Sentence length, number of words, mean
7	DESSLd	n/a	11.269	5.669	Sentence length, number of words, standard deviation
8	DESWLsy	READMSW	2.101	2.146	Word length, number of syllables, mean
9	DESWLsyd	n/a	1.266	1.240	Word length, number of syllables, standard deviation
10	DESWLlt	n/a	5.030	5.354	Word length, number of letters, mean
11	DESWLltd	n/a	3.454	3.318	Word length, number of letters, standard deviation
Text Easability Principle Component Scores					
12	PCNARz	n/a	-2.052	-1.987	Text Easability PC Narrativity, z score
13	PCNARp	n/a	2.020	2.390	Text Easability PC Narrativity, percentile
14	PCSYNz	n/a	-0.183	0.442	Text Easability PC Syntactic simplicity, z score
15	PCSYNp	n/a	42.860	67	Text Easability PC Syntactic simplicity, percentile
16	PCCNCz	n/a	0.174	-0.164	Text Easability PC Word concreteness, z score
17	PCCNCp	n/a	56.75	43.640	Text Easability PC Word concreteness, percentile
18	PCREFz	n/a	-1.608	0.887	Text Easability PC Referential cohesion, z score
Readability					
104	RDFRE	READFRE	3.969	11.364	Flesch Reading Ease
105	RDFKGL	READFKGL	18.864	15.081	Flesch-Kincaid Grade level
106	RDL2	L2	0.805	14.134	Coh-Metrix L2 Readability

Fig. 10.17 Coh-Metrix analysis results of first revision text in the Text2 column

As you can see, the RDL2 index in the Text2 column is now 14.134, just short of the suggested 15. We can improve this score further by using a guideline from Chap. 9 to bullet the lists, as shown in Example 10.5.

Example 10.5

The Process of Drug Design and Development (revised with bulleted lists)

New drug design and development is a complex process. Therefore, the process requires many people to discover, develop and successfully bring a new drug to market. This process of drug design and development must be multidimensional. It requires the combined skills of many people with different expertise. These skills include:

* medicinal chemistry;
* in vitro biology;
* drug metabolism;
* animal pharmacology;
* formulations science;
* process chemistry;
* clinical research;
* intellectual property.

Technologies that assist the overall process include:

- high-throughput screening;
- molecular modelling;
- pharmaceutical profiling;
- biomarker studies.

These technologies are also important in modern drug research.

Figure 10.18 shows the results of the revision of our example text. The RDL2 index has shot up again to 25.445.

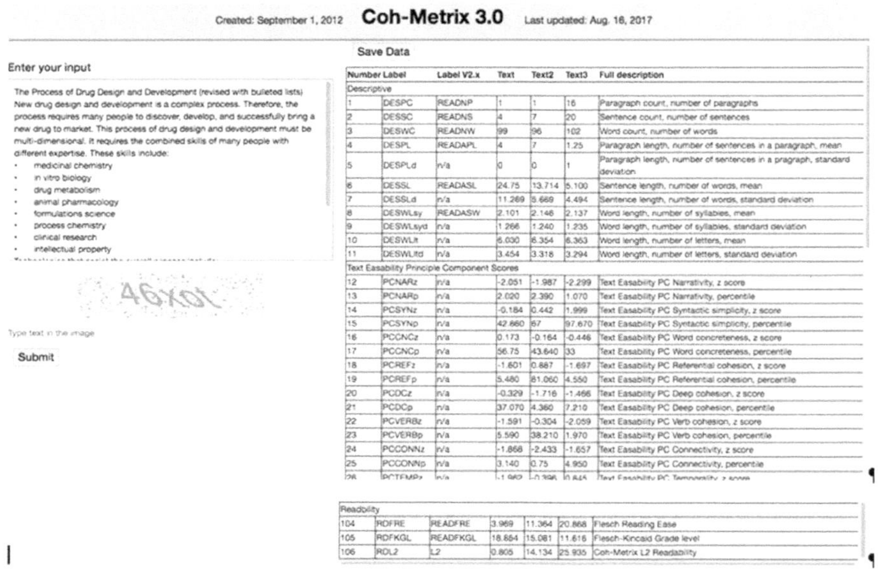

Fig. 10.18 Coh-Metrix analysis results of second revision text in the Text3 column

10.3.5   Save the Data

Clicking on the 'Save Data' button provides a $\sim$.txt file of the results. Alternatively, you can copy and paste your results directly from the screen into a word processing document. This keeps the original format intact.

A full description of all indices in the tool appears on the main Coh-Metrix website at http://cohmetrix.com/.

And finally... Even the most carefully crafted texts can present problems for L2 readers. It is often surprising—and instructive—to uncover the vocabulary, grammar or cultural issues that even fluent L2 readers encounter with reading material. For this reason, it is useful to conduct usability tests of your material with a sample of L2 readers.

Acknowledgements The authors gratefully acknowledge the creators of Compleat Web VP, Hemingway Editor and Coh-Metrix Index for their permissions to use the screenshots discussed in this chapter.

References

1. S. Vajjala, D. Meurers, On improving the accuracy of readability classification using insights from second language acquisition, in *The 7th Workshop on the Innovative Use of NLP for Building Educational Applications* (Association for Computational Linguistics, Montreal, Canada, 2012)
2. T. Cobb, *Compleat Web VP* (2019) (cited 31 January 2019), v.2, https://www.lextutor.ca/vp/comp/
3. A. Masrai, J. Milton, Measuring the contribution of academic and general vocabulary knowledge to learners' academic achievement. J. Engl. Acad. Purp. **31**, 44–57 (2018)
4. A. Long, B. Long, *Hemingway Editor*, 3.0 (2015), http://www.hemingwayapp.com
5. D.S. McNamara et al., *Coh-Metrix*, 3.0 (2013), http://www.cohmetrix.com
6. S.A. Crossley, D.B. Allen, D.S. McNamara, Text readability and intuitive simplification: a comparison of readability formulas. Read. Foreign Lang. **23**(1), 84–101 (2011)
7. S.A. Crossley, J. Greenfield, D.S. McNamara, Assessing text readability using cognitively based indices. Tesol Q. **42**(3), 475–493 (2008)